The Pilot's Radio Communications Handbook

Other books in the PRACTICAL FLYING SERIES

Light Airplane Navigation Essentials
Paul A. Craig

Handling In-Flight Emergencies
Jerry A. Eichenberger

Cockpit Resource Management: The Private Pilot's Guide
Thomas P. Turner

The Pilot's Guide to Weather Reports, Forecasts, and Flight Planning—2d Edition
Terry L. Lankford

Weather Patterns and Phenomena: A Pilot's Guide
Thomas P. Turner

Cross-Country Flying
Jerry A. Eichenberger

Avoiding Mid-Air Collisions
Shari Stamford Krause

Flying in Adverse Conditions
R. Randall Padfield

Advanced Aircraft Systems
David Lombardo

Understanding Aeronautical Charts—2d Edition
Terry T. Lankford

Aviator's Guide to Navigation—3d Edition
Donald J. Clausing

Learning to Fly Helicopters
R. Randall Padfield

ABCs of Safe Flying—3d Edition
David Frazier

Flying VFR in Marginal Weather—3d Edition
R. Randall Padfield

The Aviator's Guide to Flight Planning
Donald J. Clausing

Better Takeoffs & Landings
Michael Charles Love

The Art of Instrument Flying—3d Edition
J. R. Williams

Aviator's Guide to GPS—2d Edition
Bill Clarke

The Pilot's Radio Communications Handbook

Fifth Edition

Paul E. Illman

McGraw-Hill

New York   San Francisco   Washington, D.C.   Auckland   Bogotá
Caracas   Lisbon   London   Madrid   Mexico City   Milan
Montreal   New Delhi   San Juan   Singapore
Sydney   Tokyo   Toronto

Library of Congress Cataloging-in-Publication Data

Illman, Paul E.
    The pilot's radio communications handbook / Paul E. Illman. — 5th
ed.
      p.   cm. — (Practical flying series)
    Previously published: 4th ed. Blue Ridge Summit, PA : TAB Books,
©1994.
    Includes index.
    ISBN 0–07–031831-x (hardcover). — ISBN 0–07–031832-8 (pbk.)
    1. Radio in aeronautics. I. Title. II. Series.
TL693.I4 1998
629. 132'51—dc21   98-12759
                                            CIP

McGraw-Hill

A Division of The McGraw·Hill Companies

ISBN 0-07-031831-x (hc)
ISBN 0-07-031832-8 (pbk)

*The sponsoring editor for this book was Shelley Carr, the editing supervisor was
Scott Amerman, and the production supervisor was Tina Cameron. It was set in
Times by Joanne Morbit of McGraw-Hill's Professional Group Composition Unit,
in Hightstown, NJ.*

Printed and bound by R. R. Donnelley & Sons Company.

McGraw-Hill books are available at special quantity discounts to use as
premiums and sales promotions, or for use in corporate training programs.
For more information, please write to the Director of Special Sales,
McGraw-Hill, Professional Publishing, Two Penn Plaza, New York, NY
10121-2298. Or contact your local bookstore.

 This book is printed on recycled, acid-free paper containing a
minimum of 50 percent recycled, de-inked fiber.

Contents

Contents

Contents

Introduction

The air is filled today with pilots of all levels of experience, knowledge, and training. There are those who learned to fly at some uncontrolled single-strip airport and are still reluctant to venture too far from that uncomplicated harbor. There are those who mastered the private test at a busy controlled airport. There are the old-timers who never lost the flying bug but couldn't afford the luxury until their later years. And there are the pros: the airline captains, the executive pilots, the high-time instructors, the commercial pilots who work charters or are involved in some other money-making enterprise.

The list is hardly complete. Suffice it to say that even with exorbitant fuel and maintenance costs, there are a lot of us in the air—some good, some bad, some in between.

What's the common denominator we all share? Probably the critical one is safety—and all that safety implies. Just below safety may be the freedom that piloting your own (or rented) aircraft brings: the freedom to go places with reasonable economy and speed, and the freedom to get from here to there unencumbered by traffic lights, speed traps, and highway nuts who pass you at 85 on a 65-mph interstate. Only the pilot enjoys the freedom of space, distance, and speed.

A question, though, is the extent to which we take advantage of the benefits that the Cessna, the Cherokee, the Bonanza offer us. Said another way, how many of us find ourselves limited to the local traffic pattern or a few short hops to an uncontrolled airport because the big ones scare us? Or how many of us use—or know how to use—Approach Control, Center, or the Flight Service Stations? Or how many understand radio procedures and have mastered the techniques of pilot radio communications that make use of these various facilities possible?

Considering the emphasis placed on pilot training, medical qualifications, aircraft maintenance, operating rules and regulations, and the like, it would seem that at least a somewhat similar emphasis would be directed to pilot radio communication skills. For whatever reason, such is not the case. The literature on the subject is too sparse, the examples of radio dialogue too few, and explanations of what to say and how to say it too incomplete.

Theoretically, a book on radio communications shouldn't be necessary; the subject should be part and parcel of every pilot's training. Innumerable discussions with controllers, airline pilots, instructors, and ordinary weekend excursionists, however, indicate just the opposite. According to the pros and amateurs alike, the airwaves suffer from communications misuse, non-use, or overuse. If such is the case, there can be little doubt that a void exists—a void whether in the literature available, the pilot training process, or both.

In the effort to fill in void, this book has two primary purposes: to contribute to increased safety in flight through timely and correctly worded communications, and to equip the student and licensed pilot with the knowledge of radio communication and the various ground facilities so that his or her flight horizons are expanded beyond the local controlled or uncontrolled airport.

Designed primarily, but not exclusively, for the VFR pilot, the book discusses the whole spectrum of radio facilities and communication responsibilities. First there is multicom, where only aircraft-to-aircraft self-announce messages are exchanged. Then comes a similar treatment of unicom, followed by Flight Service Stations, Ground Control, Tower, Approach/Departure Control, and the Air Route Traffic Control Centers.

Accompanying the discussion of the various facilities are explanations of what each does, how to determine the proper frequencies, and most important, examples of what the pilot should say to contact each facility, what he should expect to hear, and how he should respond.

This fifth edition, then, is, in part, an update of its predecessors. At the same time, it includes two new chapters that may contribute to the appreciation of effective communications. One of these chapters cites recorded National Transportation Safety Board (NTSB) examples of aircraft accidents or incidents that were caused primarily by human (not equipment) radio failures. The second new chapter discusses certain general principles of communications, including why communications often fail, the dangers of "assuming" in the communication process, the role that nervousness and simple fear play in obstructing the flow, particularly in the pilot-controller relationship, and suggestions to help the pilot become a more skilled and confident communicator.

In summary, with its updates and new chapters, I hope that this edition will become a constructive addition to any pilot's library and a significant contributor to his or her storehouse of aviation-related knowledge. The ultimate objective, of course, is to help the pilot venture forth into controlled as well as uncontrolled airspaces with greater confidence and greater safety.

A Note of Explanation

With only a few exceptions, the aircraft N-numbers in the radio communication examples are those of previously author-owned airplanes. Since publication of the first edition of this book back in 1984, those aircraft have been sold to other parties—and perhaps even re-sold. Consequently, should a subsequent owner find his or her N-number cited in one of the dialogue examples, let that not be a reflection of that individual's radio skills—or lack thereof. Personal N-numbers were chosen to avoid any such implication then, now, and in the future.

Acknowledgments

In the previous editions of this book, we listed, with our sincere thanks, the many people who contributed so willingly their time to assist us in the preparation of each edition. Those involved were almost exclusively associated with the Federal Aviation Administration, but many have since transferred to other posts, retired, or have perhaps severed their relationships with the FAA. Consequently, and with the hope that they will not be offended, we have chosen not to re-name those that were so helpful in the past. It should be noted, however, that they all held responsible positions in one of these facilities:

Olathe, Kansas, Air Route Traffic Control Center

Kansas City Downtown Airport Control Tower

Kansas City International Airport Tower and Approach/Departure Control

Kansas City Flight Standards Office

Wichita, Kansas, Approach Control/Departure

McAlester, Oklahoma, Automated Flight Service Station

Columbia, Missouri, Automated Flight Service Station

National Ocean Service

Washington, DC, Airport Control Tower

Kansas City FAA Regional Office

In addition to those who previously gave so much of their time and expertise, I must now recognize the efforts of other contributors who assisted in the preparation of this fifth edition.

One is Joe Korb, a veteran of 29 years with the FAA, 20 years in various Flight Service Station positions, and currently Operations Supervisor in the Columbia, Missouri,

ACKNOWLEDGMENTS

Flight Service Station. Via several phone calls, Joe helped ensure the accuracy of various procedural matters I raised and brought me up to date on other FSS practices that had changed in major or minor ways since publication of this book's last edition back in 1993. In each conversation, Joe courteously and thoroughly answered my questions or corrected my understandings, while making it clear that rather than my impinging on his time, he was there to help the pilot in any way he could. I greatly appreciated Joe's assistance and am indebted to him for his many contributions.

Another important contributor was Col. John R. Schmidt (Ret.), of Leavenworth, Kansas. John is a former Air Force pilot who currently holds an air transport pilot rating, has well over 6,000 hour of pilot-in-command time, is a gold seal CFI and an advanced ground instructor. He is also a safety counselor for the FAA.

Again for the purposes of accuracy and completeness, I asked John to review four of the chapters with a critical eye and a ready editing pen. This he did with meticulous thoroughness, and, in the process, made several meaningful suggestions as well as factual corrections. For his efforts, I am most grateful. It was not an easy task but it was one that he accomplished with excellence.

Finally, I owe many thanks to a friend and neighbor, Chris Followell. He, at my request, reviewed the chapter on Air Route Traffic Control Centers (ARTCC) and, at the same time, answered several questions related to Approach and Departure Control. Chris is highly experienced in both areas, having served tours of duty with the FAA as an Approach/Departure controller in Wichita, Kansas, and then in Honolulu. He is currently a controller in the Kansas City Air Route Traffic Control Center, based in Olathe, Kansas. With this wealth of experience, Chris was well-qualified to critique the ARTCC chapter and ensure the currency as well as accuracy of the contents discussed in that chapter. For Chris's help and the throroughness of his review, I thank him most sincerely.

Thus to all who have contributed to this effort, past and present, I repeat my expressions of appreciation and gratitude.

The Pilot's Radio Communications Handbook

1
A Case for Communications Skills

A FEW YEARS AGO, BACK IN THE DAYS WHEN THE AIRSPACES WERE called *TCAs, ARSAs, ATAs,* and the like, a CFII friend (Certificated Flight Instructor, Instrument) was returning to Kansas City's Downtown Airport (not the city's Class B primary airport) with a student from an instrument training flight. After Center had handed the CFII off to Approach Control and the CFII had established contact, another aircraft made its initial call, also to Approach:

Pilot: Clay County TCA Approach Control, this is Cherokee November Four One Nine Six Six. Over.

Approach: *Cherokee Four One Niner Six Six, Kansas City Approach.*

Pilot: Clay County TCA Approach Control, November Four One Nine Six Six is over the interstate, and I want to land at the big airport. Over.

Approach: *Cherokee Niner Six Six, squawk zero two five two, ident, and stand by.*

Pilot: Clay County TCA Approach Control, November Four One Nine Six
Six squawking zero two five two, identing, and standing by. Over.

At this point the controller directed several other aircraft and lined the CFII up for
the instrument approach to Downtown. The controller then returned to N41966.

Approach: *Cherokee Niner Six Six, I missed your ident. Please ident again.*

Pilot: Clay County TCA Approach Control, November Four One Nine
Six Six squawking zero two five two and identing. Over.

Approach: [After a pause] *Cherokee Niner Six Six, I'm still not receiving your
ident. Remain clear of the TCA, and say your present position and
altitude.*

Pilot: Clay County TCA Approach Control, I'm still over the interstate at
three thousand five hundred feet, and I want to land at the big Kansas
City Airport. Over.

Approach: *Cherokee Niner Six Six, which interstate are you over? There are
several in this area.*

Pilot: Clay County TCA Approach Control, November Four One Nine Six
Six. I'm not sure which interstate, but it's near the city. I still want to
land at the big airport. Over.

Approach: *Cherokee Niner Six Six, I have not received your ident. Remain clear of
the TCA and stand by.*

Instead of doing what he was told, the pilot of 966 launched into an airwave-
monopolizing discourse along these lines:

Pilot: Clay County TCA Approach Control, this is November Four One Nine
Six Six. I don't know why you aren't receiving my ident. I just had it
worked on, and the mechanic told me it was fine. I've got to land at the
big airport because I told Agnes, my wife, I'd pick her and the kids up
when they got in from Chicago. What will they think if I'm not there?
Over.

Approach: *Cherokee Niner Six Six, Kansas City International is a TCA, and I
can't clear you to land unless your transponder is working. I am not
receiving your ident, so remain clear of the TCA, and please stand by.*

Continuing to ignore the explicit instructions, 966 rambled on:

Pilot: Clay County TCA Approach Control, November Four One Nine Six Six.
I just had the transponder checked because the last time I was here the
controller told me to stand by. I did, and the thing didn't work then. The
guy at the radio shop said it worked fine, but I'm still having the same
trouble. Can't you get me into the big airport? Over.

Throughout all of this, other aircraft were trying to get a word in to report positions, get clearances, and the like. But N41966 continued on and on as though he was the only one in the air.

After the last exchange, the controller saw the light. The pilot of 966, bathed in the glow of ignorance, did what he had been told. He entered "0252" in the transponder, pushed the IDENT button, and placed the switch in the STANDBY position. Of course he wasn't received!

Once the mystery was solved, the pilot, quite unabashed by his display of incompetence, was cleared into Kansas City International—the "big airport."

This is about as accurate an account of the dialogue as is possible to re-create because, of course, the instructor didn't tape the real thing. It's only one incident, and, while unusual in some respects, it's not very different from what pilots and controllers hear every day. All a pilot has to do is listen with a critical ear. Some of the garbage that filters through speaker or headset from air to ground reflects an appalling lack of knowledge that is both unfunny and potentially hazardous to the ignorant pilot and those occupying the same general airspace.

WHY THE PROBLEM?

Who's to blame for the incompetence? Oh, we could probably point a finger at the instructor who eased over the whole subject of communications, but the main thrust of accusation must be directed at the pilot himself. The pilot of N41966 obviously had little interest in the subject. Otherwise, he would have been sure that he knew what he was doing before venturing into a controlled and congested traffic area. At the same time, we can blame him for a consummate egotism that allowed him to enter such an area with so little knowledge.

Pilots such as our friend in N41966 are dangerous because they don't know what they don't know. They are the airman's example of the Peter Principle. They've risen to their level of incompetence. If the flying abilities of this character were no better than his communicating skills, I wonder if he is still among the living—assuming he continued to exercise his private pilot privileges.

Yes, we can blame the pilot for incompetence, but others also share in the blame. Let's include the CFIs and CFIIs. And let's include the literature—or lack of it—that discusses the subject of radio communications.

Without exaggeration, it's a subject that probably receives the least attention and explanation of all of a pilot's flight training. Even the material currently in print offers little guidance on what to say and how to say it. *The Aeronautical Information Manual* takes a stab at a few examples, but the examples are limited; some conclude with the almost extinct "Over."

If flight instructors (certainly not all, but entirely too many) fail to teach more than the absolute rudiments of radio procedures, there are probably three reasons:

- The instructor isn't too sure about them himself. This should be an unlikely reason, but, as just an example, when getting back into flying after an absence of several years, I was told by a young CFI always to begin calls with "This is *November* 1461 Tango" and conclude with "Over." (So that I don't leave you wondering, "November" is usually used only by controllers when the pilot calling in hasn't identified his or her aircraft type. The controller then comes back with something akin to, "November 1461 Tango, say type of aircraft." The principle: *Always* include your make of aircraft in at least the initial call-up, as "Salem Approach, Cherokee One Four Six One Tango…" "Over" is rarely used, but it can be helpful to indicate the end of a lengthy transmission. Otherwise, it's not necessary, and you almost never hear it in normal pilot-controller dialogues.)

If the instructor has accumulated most of his hours flying out of Cowslip Municipal, he probably isn't very confident of radio techniques. His own insecurity results in a superficial coverage of the subject as he preps his eager students for the FAA check rides.

- The instructor is a pro as a radio communicator, but teaching the subject takes time—mostly ground time, which is neither profitable nor exciting. So the student learns barely enough to get by.

- The airport is uncontrolled (meaning, no tower) and no controlled airport is within reasonable flight range. This is a logical reason for not teaching communications in depth, especially if the student plans to fly only on weekends and demonstrate his skills over Aunt Martha's vineyard. A good instructor emphasizes what is necessary to know, not what's nice to know.

That same instructor, however, should make it very clear that if the student (now private pilot) ever plans to fly to a controlled airport or through a Class B airspace or go on a cross-country, he must return for a thorough schooling in radio procedures. A fully equipped aircraft and a private pilot certificate entitle a pilot to land at any airport. However, the hardware and a piece of paper are hardly adequate to ensure the continued well-being of the pilot or the other airmen in the vicinity. It's dangerous to run out of knowledge—but it can happen easily and quickly to untrained pilots. The results may be devastating.

So okay—you've got a license, and you either own or rent a plane. You're a good pilot, confident of your ability. Now, like many of your counterparts, are you going to spend the rest of your flying days avoiding tower-controlled airports or being fearful of using Center (Air Route Traffic Control Center) on a VFR cross-country? If you've been well-trained in radio procedures, a busy airport or getting advisories from Center is neither a challenge nor a concern. Your radio skill makes flying just that much more fun. But if you're untrained or uncertain, you'll probably steer clear of the controlled areas and not bother Center because you think that's for the IFR pilots and the pros who wheel the wide-bodies.

This, of course, is nonsense. Admittedly, Center might not be able to help you on a busy day if you're VFR. A controller also has the right to refuse to give you routine en route advisories or track you on radar if you come across as hesitant, uncertain, or lacking in knowledge. Center controllers might not do this very often, but requests in VFR conditions have been rejected when the pilot was obviously incompetent in the basics of radio communications. Otherwise, Center exists to serve all pilots—from the greenest student to the 30,000-hour airline captain. Besides, the FAA urges us to use this as well as all other facilities available to us.

Let's be careful not to oversimplify the matter of radio procedures. Mastering them takes time and practice. To underscore that point, the FAA's *Instrument Flying Handbook* makes these comments:

> …Many students have no serious difficulty in learning basic aircraft control and radio navigation, but stumble through even the simplest radio communications. During the initial phase of training in Air Traffic Control procedures and radiotelephone techniques, some students experience difficulty.…
>
> …Communication is a two-way effort, and the controller expects you to work toward the same level of competence that he strives to achieve. Tape recordings comparing transmissions by professional pilots and inexperienced or inadequately trained general aviation pilots illustrate the need for effective radiotelephone technique. In a typical instance, an airline pilot reported his position in 5 seconds whereas a private pilot reporting over the same fix took 4 minutes to transmit essentially the same information.…The novice forgot to tune his radio properly before transmitting, interrupted other transmissions, repeated unnecessary data, forgot other essential information, requested instructions repeatedly, and created the general impression of cockpit disorganization.…

PRACTICING FOR COMPETENCE

Mastery of the technique starts with knowing what you want to say, what to listen for, how to respond, and when and how to use the mike that spreads your voice throughout the surrounding skies. As in any other field, the initial ingredient of proficiency is knowledge. The trick is to apply that knowledge in a logical sequence so that you can say what you want to say and get off the air. Once the knowledge is acquired, the next step is practice, followed by more practice, until what you know intellectually becomes an ingrained habit.

If you've ever been asked to make a speech, you know that you didn't just get up and talk. You either wrote the entire speech or outlined it, and then practiced it until you had the subject matter, sequence, body language, and voice inflections down pat. The first time around, you were probably a bit nervous. The second time was a little easier. Eventually, if you spoke or lectured enough, you became an old pro.

It's the same thing talking to the ground from an airplane. Knowledge coupled with practice will calm nerves and conquer whatever mike fright you might have. No matter how green you are, you'll come across as a professional.

You can practice in a couple of ways. One is to buy an inexpensive aircraft-band radio that picks up the various aviation frequencies. Then monitor the transmissions from your home. This method is less effective if you live far from a tower, but you can at least listen to the pilot's side of the communications exchange.

Another practice method is to use a tape recorder and do a little role-playing with yourself. You're the pilot and controller all in one. Make the initial call to Ground Control, and answer yourself as the controller would. Or pretend that you're in flight and want to land at X airport. Go through the same process. Using your knowledge of radio procedures, act out a series of scenarios on a mythical flight from the first contact with Ground Control until you have "landed," are off the active, and have called Ground Control again for taxi clearance.

Then play the tape back. Be your own worst critic. Be objective about the "dialogue." Ask yourself: "If I were a controller or another pilot listening to me, what would be my impressions of me?" If you're not satisfied, pick up the mike and go at it again.

If you practice this way enough, it won't take long to learn how to get the message across in the fewest possible words and with maximum clarity. Yes, you might feel a little silly sitting there talking to yourself, but that, too, shall pass. Even if it doesn't, it's a small price to pay for greater confidence and increased expertise.

Now you have the words down, but will you remember them, and in the proper sequence, when it comes to the real thing? If in doubt, write out what you want to say when you contact the various services. Put the notes on your knee pad and, if necessary, read from them as you make your calls. After all, people use checklists so they don't have to rely on a memory that might fail them, so why not adopt the same technique for radio contacts? (However, unlike checklists, experience makes the need for written notes unnecessary.)

A pause to regroup: I hope I am not exaggerating the case and the need for greater communication skills among the pilot population. Obviously, I don't think so. All you have to do is fly a few hours a week and keep an alert ear to what flows through the headset. On any given flight around a busy airport you'll hear everything from a terse "Okay," to a rambling recitation of superfluous trivia, to a series of mumbled incoherencies that no human or electronic decoder could decipher. If you question that statement, spend a few minutes with a controller and listen to what he or she has to say.

CONTROLLERS ARE HUMAN, TOO

Flight Service Stations, Towers, Approach and Departure Control, and the Air Route Traffic Control Centers are the pilot's valuable but unseen friends. They exist to serve the pilot and to make flying safer for all. Their services, however, aren't really free. You've paid for them through your taxes. The services are there to be used or not used, so why not take advantage of your annual donation to Uncle Sam and the taxes you find tagged onto your fuel bill?

Yes, you've paid for the service, but so has every other pilot, so it's not yours and yours alone to use or misuse. Any given service, particularly that offered by Center, can be denied you if you give an impression of incompetence. On occasion, those on the ground just don't have time to try to make sense out of nonsense and clarity out of obscurity. To do so might put someone else's life in jeopardy.

At times it might seem unlikely, but controllers happen to be human beings, too. They have good days; they have bad days. Even on their good days, though, they can quickly turn into vocal ogres when they encounter unmitigated stupidity over the air. On bad days, they can come across as halo-endowed saints when a knowledgeable pro solicits their assistance or advice.

While we're on the subject, the basic rules of courtesy over the air should always prevail. When either pilots or controllers resort to sarcasm or needless abuse, they are merely reflecting emotional immaturity—which is hardly a credit to the responsibility they bear. Controllers call it "chipping," a gentle term for "telling the other guy off."

Controllers recognize the humanity of man, and 99 percent never utter a word of recrimination when mistakes are made or ignorance shines brightly. A few don't have that level of patience, of course. They can chip with the best of them, as did one I overheard when he couldn't get a response from a pilot with whom he had just been in contact: "You gonna talk to me, boy? If you are, talk *now*."

To give them their due, controllers have to be models of tolerance and self-control to endure some of the things that go on in and over the air. Yes, some talk too rapidly, and some runtheirwordstogether so that comprehension is nigh impossible. But the performance of the vast majority, even under pressure, sets a standard of excellence in their profession that pilots should strive to match in theirs.

If you have a problem with a controller, don't let anger overrule good judgment. The radio isn't the place for chipping. Wait until you're on the ground. Then call the facility and talk to the supervisor. Explain calmly what happened. Let the supervisor take it from there. Childish spleen-venting is out of place in the adult world, whether airborne or ground-bound.

While the controller is indeed the "controller," that doesn't mean he or she has to be obeyed at all costs. You are still the pilot in command. If the controller tells you to do something that you believe might endanger you, say so. Don't follow blindly into the path of possible destruction, but don't keep the controller in the dark about your concern or your alternate action.

In a very literal sense, a team is at work: you and those on the ground. They are there to ensure your safety and that of your fellow pilots. They can fulfill their responsibilities, however, only if you keep them informed and conduct yourself with the skill expected of a licensed pilot—private or ATP.

By the same token, if you help the controller when he or she asks you to lengthen your downwind leg, make a tight pattern, land long, speed up, slow down, make a high-speed landing runout, or whatever, you'll be functioning as an effective team member. Remember: the controller can do without you, but you can't do without the controller. Whether you are new or experienced, it is entirely to your personal benefit to make it easy for the controller to do his or her job and thus help you do yours. Achieving that objective is a matter of communications—knowing *what* to say, *how* to say it, *when* to say it, and *why* it should be said. Knowledge plus skill equals professionalism.

All evidence that I have found indicates that a strong case can be made for greater pilot communication skills. The reason behind poor communication, whether it's the absence of literature on the subject or instructor reluctance to emphasize it, is secondary. The result is often a pilot's unnecessary fear of the microphone, which in turn tends to restrict his or her flying activities and limits the airborne adventures to which the license entitles him or her. The alternate result is unjustified confidence, as embodied by our friend in Cherokee November 41966.

What follows, then, will hopefully reduce any fears you might have and establish justified confidence in your ability to communicate as a professional. Whether flying is your vocation or avocation, that should be your objective.

A FEW WORDS ABOUT PHRASEOLOGY

Because we'll soon begin illustrating the various radio calls and contacts, I want to be sure that some basic phraseology principles are understood. There's nothing difficult about it, but there is a certain standardization that is both accepted and expected. Reasonable variations are, of course, permissible. The examples that follow in this book, however, generally reflect the approved wording and structure as established by the *Aeronautical Information Manual* (AIM) and the FAA's *Air Traffic Control Manual,* 7110.65, for controllers.

As to wording, aircraft N-numbers are stated individually, preceded by the aircraft type. Cherokee 1461 Tango is announced as "Cherokee One Four Six One Tango," not "Cherokee Fourteen Sixty-One Tango." Land Runway 19 is "Land Runway One Niner," not "Nineteen." "Altimeter 29.65" is "Altimeter two niner six five," not "twenty-nine sixty-five." "Heading 270" is "Heading two seven zero," not "…two seventy."

In quoting altitudes, controllers state them in terms of thousands and hundreds: "maintain three thousand five hundred" or "expect seven thousand five hundred in ten minutes." Pilots can (and do) shorten altitude quotations by saying "level at three point five" or "leaving five point five for three point zero" or "over the field at two point three." This sort of verbal shorthand is acceptable from the pilot, but does not conform to FAA standards. Consequently, you won't hear controllers using that phraseology, and in the examples I cite from now on, I attempt to conform to FAA recommendations.

Accordingly, and to be sure that you understand and employ the correct phraseology, all numbers in the simulated dialogues that follow are spelled out. "Runway 19" will appear as "Runway One Niner" because that's the way it's pronounced. "Heading 240" is stated as "Heading two four zero" and so on.

Depending on the specific reference, decimal points might or might not be included in the quotation. For instance, when citing altimeter settings, the decimal is omitted. A setting of "30.08" is communicated as "altimeter three zero zero eight." On the other hand, the decimal is included in references to radio frequencies. "Contact Ground on 121.9" is stated as "Contact Ground on one two one point niner" or "Contact Ground, point niner."

One other explanation is apropos before we get into examples. You will note that at times I use the aircraft's type and full N-number, such as "Cherokee One Four Six One Tango." On other occasions, it's "Cherokee Six One Tango." Why the difference? When making the initial contact with each controller (Ground Control, Tower, each Center sector, etc.), the type of aircraft should be identified and its full N-number given (just in case the controller is handling another "Cherokee Six One Tango," a distinct possibility in congested areas). You can shorten the call sign to "Cherokee Six One Tango" after the controller does. Once the controller abbreviates your call sign, there's no point in giving the complete identification in subsequent calls to the same controller.

Even at uncontrolled (nontower) airports, the identification process should be the same. Make the type of aircraft you're flying known to others—and hope that they extend the same courtesy to you. There's a big difference between landing behind "Zero Zero Zero Zero Alpha" and "Learjet Zero Zero Zero Zero Alpha." It would be nice to know that you're trailing a jet rather than a Piper Cub. The wake turbulence of the former can be a bit more challenging.

Yes, a fair amount of verbal shorthand is acceptable, but not necessarily correct. Despite this phraseology latitude exercised by pilots (and perhaps even tolerated by the FAA), I have chosen not to take such liberties in the communication examples. The idea is to present the correct wording and phrasing here. Accepted but unapproved abbreviations can come later, if you so choose.

2
Accident Reports and Communications Failures

WITHOUT THE BENEFIT OF ANY SPECIFIC DATA, I'VE OFTEN wondered how many aircraft accidents or near misses have been caused primarily by pilot misuse or nonuse of his or her radio communications facilities. Obviously, I'm not talking about radios that suddenly die. Instead, it's the people in the left seat who, for whatever reason, don't make known their intentions, don't listen to instructions, or don't pay any attention to what other pilots are saying and how that information might affect their own intentions. Someday, I'd like to see the results of a study that focuses solely on accidents or incidents traceable directly to this cause. The results might be surprising. In lieu of such a study right at hand, however, what follows may give some clues as to the frequency or the seriousness of the problem.

Chapter Two

THE MOST COMMON CAUSES—IN SUMMARY

The Aircraft Owners and Pilots Association's 1996 "Nall Report" indicates, once all the facts are established, that 70 to 80 percent of the general aviation accidents can be traced to human factors, while only about 8 percent are the result of mechanical or maintenance failures. "Other" or "unknown" make up the rest. If you can accept the Nall Report and then have the opportunity to review the National Transportation Safety Board (NTSB) accident summaries compiled over the past 15 years, the role that radio communications has played becomes strikingly apparent. Sometimes the role is major, sometimes it's incidental, but it's there, nonetheless. And, typically associated with so many of the NTSB accident studies are comments such as these:

- "Air/ground communications not attained."
- "Communications/information delayed."
- "Air/ground communications not used."
- "Improper interpretation of instructions."
- "Communication inattention."
- "Communications not understood."
- "Improper use of radio equipment."
- "Radio communications not maintained."
- "Failure to communicate on the Common Traffic Advisory Frequency (CTAF)."
- "Communications inadequate."

Although other factors may also have contributed to a given accident, the frequency with which some form of communicating failure is mentioned leaves at least one major impression: We need to sharpen up on our radio sending, receiving, and listening skills!

Admittedly, a mechanical failure that results in an accident is one thing; so are problems resulting from ground or aircraft avionics failures, electrical power outages, and the like. Mechanical breakdowns can, indeed, produce a few hairy moments until things get VFR and a friendly runway looms dead ahead. For the most part, though, these are hardware items, things material. While they do fail, they are amazingly dependable, if only routine maintenance is paid to their well-being. Consequently, their collective role in the accident/cause relationship is usually secondary and more often than not of little significance.

It's the human element that needs the attention, the training, the retraining, and, in tune with the subject at hand, a continuing emphasis on radio communicating skills. As pilots, one human characteristic we all share is the ability to communicate. We may not be very good communicators, but we at least are physically able to speak and to hear, and to convey and receive messages. If deficient in either of those abilities, it's most unlikely that you or I would even possess a current FAA medical certificate.

Being capable of speech and hearing, however, does not mean that, as pilots, we use these capabilities skillfully or effectively. Should you tend to question that, just listen carefully to some of the chatter that fills the air on a sunny Sunday—or any other good flying day. You'll hear all sorts of pilot-originated messages that reflect a dire need for subject organization, on-the-air brevity, clarity of message, and, in effect, plain and simple communications training. Perhaps this book, in its own way, can contribute to that so-called training.

A FEW CASE HISTORIES

To illustrate the need for training, or at least for better radio communications, let's look briefly at a few NTSB-recorded accidents, the cause(s) of same, as concluded by the NTSB, and the extent to which radio communications played a part—major or minor—in what happened. As implied earlier, these cases have been taken directly from files provided by the National Transportation Safety Board in Washington.

The NTSB has recorded, by month, over 37,000 accidents that have occurred since January 1983, up to the present. With the help of its Aviation Accident Data Specialist Analysis and Data Division, accidents in which radio communications, or lack thereof, played a part were isolated for me. Only a few (13, to be specific) are summarized here simply because of space limitations.

To protect the guilty or avoid possible embarrassment to any party, some of the details included in the NTSB printouts, such as aircraft N-numbers, dates of the incidents, or pilots' names, are not included. "Probable Causes" and "Contributing Factors," cited at the end of each case, are summaries of the NTSB's conclusions as to the cause(s) of the accident. Then, where it seems appropriate, I have added a further comment or a question. To make it easier to visualize the situations, I've also included skeletal diagrams of the various airport runway systems, based on AOPA's (Aircraft Owners and Pilots Association) 1997 *Airport Directory,* all rights reserved.

Case 1

Location: Martinsburg, WV
Aircraft: Cessna 150
Aircraft damage: Substantial
Injuries: One, minor

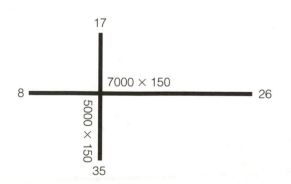

Fig. 2-1. Martinsburg, West Virginia.

The pilot left the airport without contacting the tower. He returned 30 minutes later and made several missed approaches, still without contacting the tower. Tower personnel tried to communicate with pilot and used light signals while directing other traffic. The pilot stated (later) that he saw a C-130 on the taxiway as well as several other light civil aircraft in the area. He then stated he became disoriented and lost control. Witnesses reported that the aircraft entered a steep circling maneuver which terminated when the left wing hit the ground, and the aircraft finally came to rest between runway 8/26 and the parallel taxiway. The 75-year old pilot had no previous experience in the Cessna 150, had not flown in the previous 90 days, and had never received a biennial flight review.

Probable Cause/Contributing Factors
Poor airplane handling, no radio communications, poor judgment, lack of experience in aircraft type, and several other causes, all pilot-directed.

Comment
Radio communications might not have prevented this fiasco, but at least the tower would have been in a position to help guide the pilot and prevent disorientation. It also could have saved much disruption of traffic in the airport vicinity and unnecessary directions to other pilots.

Case 2

Location: Glendale, AZ
Aircraft: Starduster II
Aircraft damage: Destroyed
Injuries: Two, serious

19

5350 × 75

Fig. 2-2. Glendale, Arizona.

1

The pilot crossed the airport at midfield and made a left turn for a close-in run downwind. Witnesses estimated he made a 180-degree descending turn to final approach. While on the final descent, the pilot observed another aircraft which was already on final slightly ahead and slightly below his aircraft. The pilot performed a hard pull-up to avoid the other aircraft and allowed his aircraft to stall without sufficient altitude to effect recovery before ground impact occurred.

Probable Causes/Contributing Factors

Pull-up excessive, airspeed not maintained; visual lookout inadequate; radio communications not maintained—pilot of other aircraft.

Comment

The NTSB summary doesn't indicate whether the Starduster was radio-equipped. If it was, whether hand-held or otherwise, this pilot was as much to blame for the near-midair collision as the pilot of "the other aircraft." If the Starduster had no radio, then the NTSB's last comment about the "pilot of other aircraft" really isn't pertinent. One can only assume, then, that the Starduster was radio-equipped and that its pilot either didn't have the set on or wasn't paying attention to advisories from other aircraft in the area. Whatever the case, radio communications could have prevented the near-miss and the eventual crash from ever happening.

Case 3

Location: Temple, TX
Aircraft: Cessna 150 and Cessna 152
Aircraft damage: C-150, substantial; C-152, destroyed
Injuries: One fatality; two serious

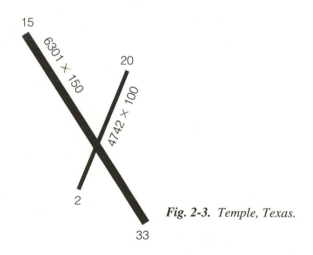

Fig. 2-3. Temple, Texas.

A midair collision occurred between a Cessna 150 in the traffic pattern and a Cessna 152 on a long, low final approach to the same runway. (*Note:* The runway number was not identified in the NTSB report.) The C-152's radio was mistuned and the C-150 had no radio installed. Consequently, neither could have heard warnings broadcast by another aircraft. At the time of the accident, the C-152 came between the C-150 and the sun during the last minute of the flight. The C-150 was hidden from the C-152's pilot's view by the left wing. No obstruction was found to account for either aircraft not seeing or avoiding the other prior to those positions.

Combined Probable Causes/Contributing Factors
Communications/navigation equipment—lack of; radio not tuned to local frequency; visual lookout inadequate; pattern procedures not followed; light conditions (sunglare).

Comment
Could two-way radios in both aircraft, tuned to the Temple Common Traffic Advisory Frequency, and properly used by the pilots have prevented this accident? We'll never know, but logic says "yes."

Case 4

Location: West Palm Beach, FL
Aircraft: Cessna 403
Aircraft damage: None
Injuries: None

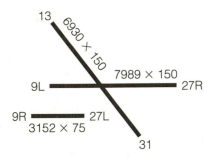

Fig. 2-4. West Palm Beach, Florida.

A Learjet 35 was cleared onto Runway 9L and held. The Learjet crew confirmed with the tower personnel that the transmission was intended for them and taxied into position and held. The tower then cleared a Cessna 403 into "position and hold" on Runway 13, and cleared the Learjet for departure. The Learjet acknowledged the call and the clearance. Both aircraft then began to take off and missed each other by about 10 feet after becoming airborne. The Cessna pilot stated later that he thought the take-off clearance was for him.

Probable Cause and Comment
The Cessna pilot did not acknowledge tower instructions and obviously did not listen to instructions to the other aircraft, the Learjet. (In later chapters, I'll emphasize, a lot more than once, the importance of terse but complete acknowledgments of ATC [Air Traffic Control] instructions or information.) This was a pure case of radio communications fail-

ures on the part of the Cessna, with only 10 feet saving the combined passenger count of 11 occupants from almost certain injury or death.

Case 5

Location: Tehachapi, CA
Aircraft: Cessna 206 and USAF T-38
Aircraft damage: Both destroyed
Injuries: Fatalities, four

A USAF T-38 aircraft and a Cessna 206 collided in midair during visual flight operations in a Military Operations Area (MOA). Radar data revealed that the overtaking T-38 collided with the Cessna from the right. The Cessna was on a local government contract terrain photo mission and had almost completed a left 360-degree turn. The T-38 was transiting the MOA in connection with a military mission. Both aircraft were operating under VFR and were not in contact with the MOA controllers. Neither pilot was required to establish or maintain positive radio or radar contact with the air traffic control facility responsible for the area. However, *AIM*, the *Aeronautical Information Manual*, states: "Pilots operating under VFR should exercise extreme caution while flying within an MOA…Prior to entering an active MOA, pilots should contact the controlling agency for traffic advisories." Radar and radio assistance was available to either pilot. The T-38 was using a transponder with Mode C readout. The collision occurred at about 8,700 feet msl (mean sea level).

Probable Cause/Contributing Factors

Visual lookout by both pilots was inadequate; procedures relative to operating in an MOA were not followed; available radio communications and radar assistance were not used by either pilot.

Comments

Flying in any active MOA when military training flights are in progress is playing Russian roulette. Radio contact with a Flight Service Station or the responsible Air Route Traffic Control Center before entering an MOA, while not an FAA requirement, should be the standard practice for any civilian pilot. More details on this will be given later, though, because the subject is too important to dismiss with just a casual warning.

Case 6

Location: Ruston, LA
Aircraft: Cessna 310; Cessna 152
Aircraft damage: Cessna 310, minor; Cessna 152, substantial
Injuries: Three uninjured

Both airplanes were in the traffic pattern. The C-152 made radio calls in the blind; the C-310 did not. Both planes were on final at the same time, with the C-152 in the lead and the C-310 lower. The C-310 passed the C-152 just before touchdown and was on the

18

5000 × 100

36

Fig. 2-5. *Ruston, Louisiana.*

ground as the C-152 rounded out right behind it. The C-152 pilot then saw the C-310, added full power, and pulled up. The left horizontal stabilizer of the C-152 struck the C-310's vertical stablizer. The C-310 stopped and the C-152 landed in front of the C-310 on the same runway.

Probable Causes/Contributing Factors

Standard traffic pattern procedures not followed, inadequate visual scanning of the pattern area, and lack of radio communications by the Cessna 310.

Comment

The accident report does not state whether the C-310 had its radio on. If it did, the pilot(s) was obviously not paying attention or he (she) would have heard the C-152's various calls. This again appears to be an incident primarily caused by radio communications failures, compounded by visual inattention and nonadherence to traffic pattern procedures.

Case 7

Location: Zephyrhills, FL
Aircraft: Cessna 152
Aircraft damage: Substantial
Injuries: One uninjured

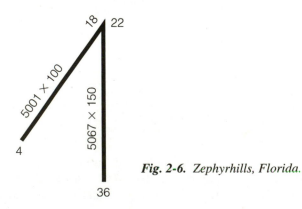

18   22

5001 × 100

5067 × 150

4

36

Fig. 2-6. *Zephyrhills, Florida.*

The pilot stated that when he contacted unicom, he heard Runway 32 was active, but since it did not exist, he interpreted the active as Runway 22. After touchdown with a right quartering tail wind, the aircraft bounced, drifted off the runway, flew over a ditch, stalled, then touched down in sand and nosed over. The unicom operator stated that he heard the pilot call for a runway advisory, but before he could issue an advisory, the pilot responded, "Understand Runway 22 Zephyrhills." The operator then advised that the winds were favoring Runway 36, but there was no response. The pilot also stated that he did not fly over the field to observe the windsock.

Probable Causes and Comments

The student pilot incorrectly interpreted the favored runway information and then either did not hear or disregarded the advisory about the winds and the favoring of Runway 36. After landing on the wrong runway, the damage to the aircraft was primarily the result of improperly recovering from a bounced landing, the tail wind conditions, and the pilot's lack of experience.

Case 8

Location: Cove Neck, NY
Aircraft: Boeing 707-321B (Avianca Airlines)
Aircraft damage: Destroyed
Injuries: 73 fatal, 81 serious, one minor

On January 25, 1990, at approximately 2134 EST, Avianca Airlines Flight 052 (AVA052) crashed in a wooded residential area in Cove Neck, Long Island, New York. The flight was a scheduled international passenger flight from Bogota, Colombia, to John F. Kennedy International Airport, NY, with an intermediate stop at Jose Maria Cordova Airport, near Medellin, Colombia. Because of poor weather conditions in the northeastern part of the United States, the flight was placed in holding three times by ATC, for a total of one hour and 17 minutes. During the third period of holding, the flight crew reported that the aircraft could not hold longer than five minutes, that it was running out of fuel, and that it could not reach its alternate airport, Logan International, Boston. While trying to return to the airport (JFK), the aircraft experienced a loss of power to all four engines and crashed approximately 16 miles from the airport.

Probable Cause

The failure of the flight crew to manage the plane's fuel load adequately and their failure to communicate an emergency fuel situation to ATC before fuel exhaustion occurred. Contributing to the accident was the flight crew's failure to use an airline operational control dispatch system to assist them during the international flight into a high-density airport in poor weather. Also contributing was inadequate traffic flow management by the FAA and the lack of standardized, understandable terminology for pilots and controllers for minimum and emergency fuel states. The NTSB also determined that windshear, crew fatigue, and stress were factors that led to the unsuccessful completion of the first approach (at JFK) and thus contributed to the accident.

Comments

Not clearly stated in the above NTSB report were factors such as the possibility of a language barrier problem and the failure, for whatever reason, of the AVA 052 crew to declare clearly and forcefully the existence of a bona fide emergency. It has been reported that had ATC been aware of the dire fuel shortage situation, the aircraft would have been given landing priority. Other factors contributed to the accident, but it all might have been avoided had radio communications, particularly from the aircraft, been clearer, more forceful, and more descriptive of the true emergency the flight was facing.

This is a commercial airline example of communicating difficulties, but it's included here to show that not even the professionals are immune from falling prey to confusing, conflicting, or misleading messages.

Case 9

Location: Statesboro, GA
Aircraft: Taylorcraft; Cessna 150
Aircraft damage: Both substantial
Injuries: Three uninjured

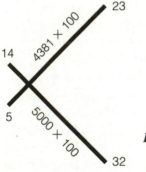

Fig. 2-7. Statesboro, Georgia.

The Taylorcraft entered a standard traffic pattern for Runway 14 at about the same time the Cessna 150 was maneuvered for a straight-in landing on Runway 5. Subsequently, the two aircraft converged and collided at the intersection of the two runways. The Taylorcraft pilot said he didn't see the Cessna before impact. The C-150 pilot obtained an airport advisory from unicom and was told of work on the end of Runway 14. He said he was also told that Runway 5 was the active runway and that there was no other reported traffic in the area. Also, the C-150 pilot said that he made another advisory call on short final approach, but did not hear any advisory calls from the Taylorcraft. The C-150 pilot saw the Taylorcraft just before impact and attempted to avoid a collision by initiating a go-around. The Cessna, however, did not gain enough altitude and it struck the Taylorcraft. The *AIM* states that standard traffic pattern entry at uncontrolled airports is from a 45-degree entry to the downwind leg. *AIM* also recommends inbound traffic advisories be made 10 miles out and entering downwind, base, and final approach.

Probable Cause

Failure of both pilots to perform a visual lookout sufficiently adequate to avoid a collision at an uncontrolled airport. A related factor was the failure of the Taylorcraft to communicate on the Common Traffic Advisory (CTAF) and to provide inbound advisories.

Comment

Each pilot was right and wrong in this case. The Taylorcraft made the proper entry into the traffic pattern, but failed, for whatever reason, to communicate his or her position and intentions. The Cessna pilot made the proper final approach calls but violated the recommended pattern entry and pattern procedures. Fundamentally, however, if the two aircraft had been in radio communication, or had known where each other was and the intentions of each, the accident probably would never have occurred.

Case 10

Location: Old Bridge, NJ
Aircraft: Cessna 172; Piper PA-28-180
Aircraft damage: Both, substantial
Injuries: Five uninjured

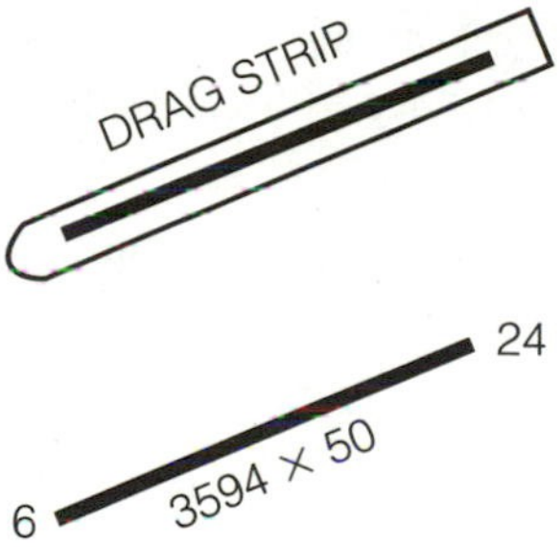

Fig. 2-8. Old Bridge, New Jersey.

Runway 06 has a displaced landing threshold because of tall trees at the approach end. The Piper was landing as the Cessna was starting its takeoff roll. The two planes collided on the runway. Neither pilot heard radio position reports from the other pilot. This is an uncontrolled airport.

Probable Cause

The failure of both pilots to maintain proper visual lookout. Related to the accident was inadequate radio communications by the pilots.

Comment

Even more at fault than the lack of visual lookout, in my opinion, was the apparent absence of any radio communications. Had both aircraft had their radios on, tuned to the proper CTAF, and both pilots communicated their positions and intentions, this is another accident that probably wouldn't have happened.

Case 11

Location: Pittsfield, MA
Aircraft: Piper PA-28-180; Piper PA-38-112
Aircraft damage: Both, substantial
Injuries: One minor, four uninjured

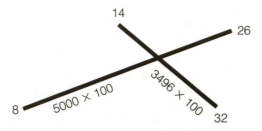

Fig. 2-9. Pittsfield, Massachusetts.

Both aircraft were on final approach to Runway 26 when they collided about one-quarter of a mile from the runway threshold and about 200 feet above the ground. Both pilots stated that they reported their positions on unicom. Several witnesses, however, who were monitoring the radio, stated that they only heard the PA-38 pilot report her position. The radios of the PA-28 were found tuned to the unicom frequency of its departure airport.

Probable Cause
The failure of the PA-28 pilot to tune his radios to the proper frequency and his failure to see and avoid the PA-38.

Comment
This seems to be pretty cut and dry: The radio is a great instrument, but it's next to worthless when it's not tuned to the right frequency. Just another example of an accident that probably should never have happened.

Case 12

Location: Conroe, TX, (Montgomery County)
Aircraft: Bellanca 17-30A; Piper PA-32-300
Aircraft damage: Bellanca, destroyed; Piper, substantial
Injuries: One, serious; one, uninjured

The right main landing gear of the Bellanca landing on Runway 19 struck the windshield and engine cowling of the Piper on a takeoff roll on Runway 14. Both planes were equipped with two-way radios and both pilots reported that they had announced their intentions on the manned unicom on 122.95. The VHF radio on the departing Piper, however, was found tuned to 122.7. The unicom operator did not recall hearing either pilot announce his intentions prior to the accident. The winds at the time of the accident were reported from 120 degrees at 8 knots, and Runway 14 was in use at the time.

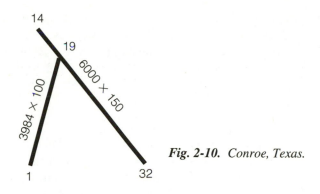

Fig. 2-10. *Conroe, Texas.*

Probable Cause

The failure of both pilots to maintain visual lookout. Contributing factors were inadequate radio communications by both pilots and the failure of both to follow procedures and directives.

Comment

More than anything, this appears to be primarily a case of radio communications. Yes, the favored runway was 14, and the Bellanca landing on Runway 19 created the potentially dangerous cross-traffic situation. The incorrect tuning of the Piper's radio to 122.7, however, made it impossible for any transmission to be heard locally. By the same token, the Piper, not being tuned to 122.95, couldn't receive any transmission the Bellanca might have originated. Operating thus in a communications vacuum, a cross-traffic meeting of aircraft was invited. Coupled, in this case, with apparent inattention to the possibility of other traffic in the area, the collision was almost fore-ordained.

Case 13

Location: Quincy, IL
Aircraft: Beech 1900C; Beech King Air
Aicraft damage: Both destroyed
Injuries: 14 fatal

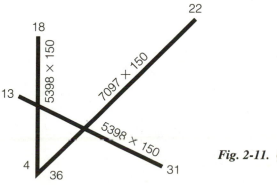

Fig. 2-11. *Quincy, Illinois.*

Chapter Two

A United Express Beechcraft 1900C, Flight 5925, collided with a Beechcraft King Air A90 at Quincy Municipal Airport, near Quincy, IL. The United Express flight was completing its landing roll on Runway 13 and the King Air was departing on Runway 4. Both pilots and 10 passengers on the 1900C and both pilots on the King Air were killed. The 1900C was on an IFR flight plan and operating under FAR Part 135. The A90 had not filed a flight plan and was operating under FAR Part 91.

Comment

At the time of writing, the NTSB had not completed, or at least published, its final analysis and probable cause(s) of the accident. With no knowledge of the accident other than what has already been printed in the various media, it would nonetheless seem most likely that a radio communications failure played a significant, if not exclusive, causal role. With experienced pilots in the cockpits of both aircraft, it is hard to believe that the King Air would have attempted a cross-traffic takeoff had he been aware of the landing United Express. And it's equally hard to believe that the United pilot would have continued an apparent normal post-landing rollout had she known that the King Air was taking off on a heading that would cross United's path at almost a 90-degree angle. For whatever reason, a breakdown in radio communications, either transmitting or receiving, by one pilot or the other, would appear to have been a major contributing factor to this accident.

Perhaps by now you've had enough of accident-summarizing. Unfortunately, though, these are only a handful of literally thousands of incidents, ranging from near-misses up to total aircraft destruction, that have been influenced by pilot misuse or non-use of radio communications. But just the few examples, picked at random from the NTSB files, would hopefully convince even the most dubious of the role the radio plays in making general aviation a safer source of pleasure, travel, and, yes, for some, a means of livelihood.

That said, let's talk next about the real meaning of "communications" and some of the barriers that most of us have to overcome if we are to get our messages across clearly, tersely, and accurately.

3
Breaking the Communications Barriers

THE EXAMPLES OF RADIO OVERUSE, MISUSE, OR NONUSE IN CHAPS. 1 and 2 should make it reasonably evident that there is a communications problem—or, put another way, a problem in communicating. Barriers, whether conscious or unconscious, do exist which can often affect our ability to get the message across or to grasp fully the message we're supposed to be receiving. And, unfortunately, many of these barriers exist in families, businesses, social relations, politics, international relations, and, yes, between pilots and ATC ("air traffic control" or "air traffic controllers) or other ground radio facilities. The ability to communicate in such ways that unfailingly produce mutual understanding between sender and receiver seems to be an elusive human talent.

That talent, if lacking, quite obviously can't be developed in these few pages. If we're aware of some of the common barriers, however, particularly those that relate to pilots and pilot communications, there is no reason why each of us can't become highly professional in the use of the radio, regardless of actual flying experience. As the few accident cases cited in Chap. 2 illustrate, this matter of effective communications is important stuff. Done well or badly in the aviation environment, unlike most situations on the ground, it can mean the difference between life and death.

COMMUNICATIONS DEFINED

The word *communicate* comes from the Latin *communicare,* meaning "to share; to make common; to make known." *Communication,* then, is the process by which something is made known, shared, made common.

In essence, if, between two people (or, carried to the extreme, between two nations), there is no commonality, no sharing, there really has been no communication. That doesn't mean that words haven't been exchanged. In a person-to-person situation, I may have heard the sound of your voice, but just hearing is not the same as understanding or correctly interpreting what you said. It's only when I have mentally grasped the meaning behind your words that there has been communication in the truly literal sense. I may not agree with your message, but that doesn't alter the fact that I *understand* it, regardless of how strongly I might reject, resist, or resent what you're saying.

Another way of looking at *understanding* takes us back to Webster's Dictionary and the definition of *comprehend.* Roughly defined, *comprehend* means to "grasp mentally, to take hold of with the mind." *Understanding,* then, is just a synonym for *comprehending,* for mentally grasping what is being said, the information being conveyed, the directions being issued. Until the listener fully comprehends what the sender wants the listener to know or do, there has been no fruitful communication. Again, there has been noise, in the sense of voice sounds between the parties but no total sharing, commonality, nor comprehension. The barriers, one or several, have been blocking the full and meaningful flow of communications.

SOME OF THE BARRIERS THAT AFFECT COMMUNICATIONS

Improving one's comunicating skills begins, of course, by becoming familiar with these barriers and how they affect the transfer or receipt of information. In normal interpersonal relations, there may be many, but some, such as body language, physical appearance, dilution or distortion of messages as they are passed from one party to another, or mistrust of the other party, aren't really issues in the pilot radio communications process. Those that do play a role, however, include the following.

Wandering Attention

This is something we've all experienced, be it in the classroom, at a business meeting, in church, with a spouse or children, driving a car, or in the air. As a pilot, have you ever been flying along on some crisp and cloudless day in sort of a dreamland, thinking about things that have no immediate relevancy to what you're doing? The radios are on, tuned to the proper frequency, the traffic is light, the visibility unlimited, the scenery outstanding, and there you are, a million mental miles away as you enjoy flight at its best.

Meanwhile, the radio chatters in the background, but the chatter falls on deaf ears. Whether through speaker or headset, you hear sounds but they're only meaningless sounds to a wandering mind. You're *hearing* but you're not *listening*. While you're off in outer space, however, another pilot is reporting his position and altitude to ATC or to a unicom station, and that report locates his or her aircraft squarely in your flight path. Might it not be nice to be aware of that potential obstacle? One would think so, but with attention drawn elsewhere, the sounds from the radio hit your eardrums and go no further. The message never gets through because you were only hearing—not listening.

And that's hard work—that process called *listening*. It requires your full attention, your ongoing analysis of the information and how that information could, would, or might affect you and your intentions. If you learn that your safety is threatened, what defensive actions should you take? What radio reports should you initiate if any? and so on. Whether it's a weather warning, a traffic report, an ATC instruction to another plane in your general vicinity, or a call to you from a controller, every radio exchange has the potential of containing information important to you. Maybe it's only nice-to-know information, but just one word or one brief transmission might materially affect your flight plan or what you do next.

The point is presumably obvious: Pay attention, pilots. Keep your radio turned up. Develop a second hearing sense so that even subconsciously you catch your aircraft N-number when ATC calls. Learn to quickly grasp the gist of transmissions to and from other aircraft and the possible importance of those transmissions to you. Start actively *listening* as soon as a voice interrupts a momentary radio silence. Especially, be on guard against wandering attention. It can be dangerous to your well-being.

If you want some examples of wandering attention or inattention, go back to Chap. 2 and Cases 3, 11, and 12 in which the pilots had tuned in the wrong frequencies. Or Case 4, where the Cessna pilot *thought* the tower's takeoff clearance call to the Learjet was for him, despite the fact that the Learjet pilot had acknowledged the call and the clearance. The result: a midair collision was avoided by only 10 feet. The Cessna pilot had obviously not been monitoring and listening to his radio…which leads to another barrier—one that is both common and dangerous, especially in an airplane.

Assumptions

The Cessna pilot *assumed* that the clearance call was for him; the pilots with the mistuned radios, in addition to not paying attention to the settings, *assumed* they were transmitting on the proper frequencies. The pilot in Case 7 heard Runway 32 from unicom, but since Runway 32 did not exist at the Zephyrhills airport, he *assumed* unicom meant Runway 22—and then finally came to a stop upside down because of winds favoring another runway and his inexperience in handling crosswinds.

This matter of assuming, of taking things for granted, can get us into lots of trouble, as these and countless other pilots have learned. Factors that cause the receiver or listener to fall prey to making assumptions include, among many, unfamiliar words or expressions (as company jargon, technical terms, acronyms, big words [often used to impress]); words or expressions that arouse anger, disagreement, arguments; a tone of voice that could imply ignorance, inability, or stupidity on the part of the receiver. Individually or in combination, these factors can cause the listener to make certain assumptions about what is being said. Then, based on the assumptions, the listener either stops listening to anything the speaker says or he or she acts on the basis of whatever assumptions he or she has reached.

How can you minimize assumptions and maximize clarifications? We'll come to that because, in part, that's the very purpose of this book. For now, though, let's stick to the barriers that so often adversely affect communications.

Thinking versus Speaking Speeds

This is another of those human factors that leads to both wandering attention and assuming. You may see slightly different figures quoted, but people generally think at the rate of about 400 words per minute and talk in the 125- to 150-words-a-minute range. That difference lets the listener jump ahead of the speaker, thus encouraging him or her to conclude what the speaker is going to say well before the actual words are uttered. In effect, the listener is saying, "Okay, okay, let's get on with it. I know what's coming." Unfortunately, though, during this period of anticipating, the listener stops listening—perhaps to his or her eventual regret.

In your own experiences, have you ever attended a meeting on, say, a new procedure or a better way of doing something, but as the talk droned on, you found yourself mentally jumping ahead of the speaker, certain in your own mind that you knew what he or she was eventually going to say? And then, perhaps some time later, did you discover that you didn't have the whole picture at all?

Maybe it was a boring subject that created wandering attention and stifled listening; maybe you assumed that you understood what was expected of you; or maybe it was your ability to think faster than the speaker could talk. Whether it was one or a combination of causes, the fact remains that listening stopped. Whatever else is said, important though it might be, falls on tuned-out ears. "I've got the message, so let's move on," say we.

But did we get the message? Do we have the whole picture? During this nonlistening period, isn't it possible that the speaker might have come up with some additional information or instructions that would change the whole context of the message? Maybe; maybe not—but possibility is what makes wandering attention, assuming, and the thinking-speed factor so dangerous.

For the pilot, keep this thinking/speaking speed ratio in mind when you become the sender, the transmitter. Before keying the mike to make a call, plan what you're going to say. Next, be sure the air is clear and no one else is talking. Then say what you have to say, but say it tersely, distinctly, and with only enough words to get your message across. Remember that with whomever you are communicating—be it a ground controller, a unicom operator, a flight service station specialist—or when making position reports—as at nontower airports—each listener, once you get started, will rather quickly assume what you're going to say. So don't drag out the obvious; don't dawdle; don't mumble; don't ramble. Don't be like the Cherokee 41966 pilot back in Chap. 1!

In essence, keep the Navy's admonition to writers and speakers in mind: KISS, Keep It Simple, Stupid.

Semantics

Here's another barrier that can make communications confusing, particularly for new pilots, relatively inexperienced pilots, or pilots not familiar with the landmarks or the local procedures in the area in which they're flying. Combine semantics—the meaning of words—with the jargon, the slang, and the acronyms associated with aviation and you have the ingredients for all sorts of miscommunication.

One of the things that makes English such a difficult language to learn is the variety of meanings so often given to a single word. For instance, how many different ways could you use or define such everyday words as *machine, sink, quarter, valve, face, land, pocket?* The list of similar multimeaning examples is almost endless. Just check any dictionary.

Then to complicate matters further for the pilot, throw in the abbreviations, the acronyms, the verbal shorthand so common in the language of aviation. Do that and the conditions are ripe for a fair degree of interpersonal misunderstanding. It's almost a new language, with some new words, new phrases, and many new meanings to familiar words. Take, for example, *squawk, squawk standby, fly the final, drag, lift, flap, chord line, dihedral, stalls, Contact ground point niner, position and hold, EFAS* (Enroute Flight Advisory Service), *ATIS* (Airport Terminal Information Service), *TAC* (Terminal Area Chart), *empennage, behind the power curve,* etc.—this is another of those almost endless lists.

From a slightly different point of view, language probably played a critical role in the Avianca Airlines accident (Case 8, Chap. 2). Suspected but not proven is the belief that the Avianca pilot, not thoroughly familiar with English and all of the standard international aviation phraseologies, felt that advising ATC of an acute fuel shortage was sufficient to grant him an immediate landing clearance. Consequently, he never literally declared an emergency. Had he done so, that would have alerted ATC to the real situation, followed by the clearance he so desperately needed. In essence, this was apparently a problem of both language and semantics, culminating in an unfortunate and probably avoidable accident.

There's no way to eliminate the semantic barriers or the in-house jargon you hear in every company, every industry, every profession. Each has its own written and spoken language that every newcomer must master if he or she is to succeed in that environment. But until mastery emerges, the opportunities for misunderstandings are myriad.

Filters

Most messages have two parts: the *content* part, meaning a statement of the idea, whatever that "idea" may be, and the *emotion* part—the feelings the idea itself might arouse, plus *how* the content part is delivered, the words used, the tone of voice, the way the speaker emphasizes or stresses the various words. At the same time and equally important is the receiver's emotional reaction to the message. For example, a tower controller calls Cherokee 1234 Alpha with this instruction:

"Cherokee 1234 Alpha, turn right now."

As a simple request or instruction, with no sense of urgency about it, these few words would probably arouse no emotional reaction on the part of the pilot. Would that calm reaction prevail if the controller said the same thing this way?

"Cherokee 1234 Alpha, *turn right NOW.*"

Probably not. Voice volume and emphasis would imply an emergency situation that demanded immediate action. Same words, same instruction, same idea, but the emotion behind the words conveys an urgency not necessarily evident in the basic message itself.

Here is another example where word emphasis can produce six entirely different meanings to a brief six-word sentence. (Put particular stress or emphasis on the italicized words.)

I never said he stole chickens.

I *never* said he stole chickens.

I never *said* he stole chickens.

I never said *he* stole chickens.

I never said he *stole* chickens.

I never said he stole *chickens.*

This is just a simple illustration, but it rather effectively demonstrates how oral or written emphasis on a word, a phrase, or even on a whole sentence can drastically alter the meaning of a message.

Perhaps Fig. 3-1 will clarify these issues of message content, emotions, and filters. As is apparent, stage (1) represents the content, the idea, what the speaker thought he or she said. During the verbalizing, however, the message itself can be filtered by a distraction, noise, excitement, anger, irritation, humor, fear on the part of the speaker, or any other distorting influence. So no matter what the speaker thought he or she said, what

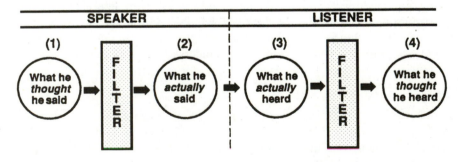

Fig. 3-1. The "filters" that distort messages make it a long way from (1) to (4).

passes through the so-called filter and emerges as what was actually said [stage 2] may not be the message the sender intended.

The listener, meanwhile, in stage (3), hears what was actually said, including the idea itself and to whatever extent it was distorted by the speaker's filter. But now the listener's filter enters the picture. What the listener actually heard may be further distorted, positively or negatively, by his or her own emotional reactions to the message, by how the message and its component parts are interpreted, by the listener's feelings of anger, frustration, acceptance, confidence, degree of trust in the sender, fatigue, mental alertness, and so on. Whatever role the listener's filter plays in the message-reception process, in stage (4) out comes what the listener *thought* he or she heard. It is on this basis, then, that the listener responds.

Suffice it to say that it's a long way from stages (1) to (4). It's thus a journey potentially fraught with misunderstandings, miscomprehensions, and mistakes. Must it be a rough journey, though? No, as I hope will become evident as we move along here and in future chapters.

Fear

This has the potential of being a serious communication barrier for any pilot but probably more so for the new pilot or the relatively low-time weekend flier. It's not just the language of aviation that may be somewhat foreign; there's also the very real concern about getting on the air and saying the wrong thing or sounding sort of stupid. If a fear exists, it's entirely understandable, because, among other things, what you say over an open mike is heard by every other pilot who is within range and tuned to that frequency. That factor alone can be intimidating, especially for the inexperienced pilot.

Reduced to the simplest, two broad fears, or certainly concerns, stand out. The first is the fear of talking and communicating over the mike, while the second is that of being the receiver or listener and having the responsibility to understand and comply with whatever instructions are issued.

As I've already indicated, the most common fear as sender or transmitter is that of making mistakes or of sounding incompetent, confused, uncertain, or disorganized. Very natural or normal, this is a fear that can be put to rest partially by knowledge—by knowing what to say, how to say it, why you should say it, and when to say it.

Knowledge alone, however, isn't enough. Even with the basics firmly in mind, the second step in overcoming whatever fear may exist is developing the necessary sending/transmitting skills. Knowledge is fundamentally factual. It's *what* you're supposed to do or say; the skills are *how* that knowledge is applied. The former comes from study and mental absorbtion; the latter is the product of practice and hands-on experience.

To begin developing the necessary sending skills, as suggested in Chap. 1, think about all the radio transmissions you would make in a normal flight, from the first departure call on the ramp to the last call after landing. Write down the dialogue, if necessary. Then, whether in front of a mirror or while driving down the highway, rehearse each call until you have the sequence and flow of information clearly in mind. With frequent rehearsals or practice sessions, it won't take long to develop the level of radio expertise that every pilot should be seeking.

The second broad fear common to pilots is understanding what a controller is telling them. Instructors hear it all the time from their students: "But I just can't understand what she wants me to do!" Another complaint is that he never heard the tower call his N-number, while a third, turning to his instructor, asks, "What did she (the controller) say?"

Overcoming the "understanding" problem really begins with having at least a general idea of the type of advice, information, or directions a controller might communicate to you, depending on where you are or what you might be doing. By that, I mean: Are you ready to taxi out for takeoff? Do you want approval to take off? Are you approaching a tower-controlled airport, contacting a Center for traffic advisories, or calling a flight service station specialist for a weather advisory? Obviously, you can't guess exactly what the controller will specifically say. With a little experience, however, and helped by the many examples cited in the pages that follow, you'll get ideas rather quickly about what information you can expect and the sequence in which it will be given. Then with study and attentive monitoring of other calls while in the air, it won't take long for you to feel quite comfortable about being able to understand and respond to advice or directions that come through your headset.

Despite what I just said, this is an area of radio communications for which there is no neat formula that guarantees immediate success. Instead, quickly recognizing and responding to calls addressed to your aircraft and understanding what is being told to you are skills that come with time and experience. The odds are good, though, that after only a few hours in the air you'll find your overall receiving ability has sharpened to the extent that it is no longer a source of concern or fear.

Now, all you have to do is master the transmitting part and you'll have the whole challenge of radio communications down pat. Right? No…not completely right, because there's one other thing to consider…

WHAT IF YOU DON'T UNDERSTAND?

No matter how many hours you have in your log book, there will be times when you just plain won't or can't understand what a controller is telling you. Some controllers speak rapidly; some occasionally slur their words; someruntheirwordstogether so that comprehension is nigh impossible; some may use a term with which you're not familiar; perhaps your radio reception is fuzzy; maybe somebody cut in at the very moment ATC is telling you what to do and all you hear are squeaks and squawks. Whatever the case, you don't get the controller's message.

Here is where uncertainty can have unpleasant consequences. Above all, if you haven't understood, don't just "Roger" the instruction and then pray that whatever you do will be the right thing. Remember that *communication* means *sharing, making common*. If you and the controller haven't shared a common message, there has been no communication.

The trouble is that people are often reluctant to admit that they didn't understand a directive or an instruction—especially if they feel that instant comprehension is expected. It's a matter of preserving one's self-esteem, of not losing face. One of the most meaningless questions an instructor, a boss, or a parent can ask is, "Do you understand?" Unless complete trust between the two parties exists, not many people are willing to say, "No. I didn't understand one thing you said." That's admitting ignorance or, worse, stupidity.

Pilot and controller talk to each other on a one-on-one basis, but there's still that unseen audience out there capturing everything the two are saying. Well aware of this, pilots uncertain about their radio communication skills face the added concern of exposing that uncertainty to their airborne compatriots. So he or she dutifully "Rogers" the instructions, hoping, with fingers crossed, that everything works out all right. Maybe it will; maybe it won't. Personally, though, I don't think playing Russian roulette a few thousand feet in the air is a very sound practice.

If you have learned what you're supposed to say, if you've practiced and rehearsed the various calls, if you've carefully monitored and listened to exchanges between other pilots and controllers, confidence in your communicating abilities should be growing rapidly. Even then, however, expect to run into occasions when you just don't or can't understand what someone on the ground has told you. When that occurs, immediately ask for clarification:

"Tower, Cherokee Three Four Alpha, say again."

"Tower, Three Four Alpha, say again more slowly."

"Your transmission was garbled. Say again, please."

"Am unfamiliar with the term. Please explain."

"You were cut out. Please repeat instructions."

"Am unfamiliar with the area. Please identify location of reporting point." (The tower told you to report "over the twin stacks," but where are the twin stacks?)

"Tower, did you say *right* downwind?" (The normal pattern is left downwind. Did you understand the tower correctly?)

Believe me. Every controller would much rather have you clarify an instruction than go off on some tangent that could endanger you or other aircraft operating in the same airspace. Covering up uncertainty just to save face is hardly worth the potential consequences.

The controller is similar to a coach, with the pilots on the ground or in the air as the players. The controller is calling the plays. He or she is in charge. The players are expected to do what he or she says. If a "play" won't work, the "coach" should be told. If there's misunderstanding or confusion, it had better be cleared up *now,* because the "game" in the air is far more consequential than any earthbound contest.

CONCLUSION

Yes, as I hope is apparent, a "team" relationship does exist between the pilot and ATC. What complicates the scenario, however, is that there is no opportunity to establish a person-to-person relationship. There's no opportunity to *see* the other person, to observe his or her facial expressions, gestures, or physical behaviors that constitute the "silent" language, the body language that so often communicates messages far more effectively than the mere transfer of words themselves. Assuming a properly functioning radio, of course, it's strictly a speaking–listening relationship that is totally dependent upon the clarity of the communications between the two parties. When there is comprehension, there is harmony; when there is harmony, each party is a winner.

At the same time, let's not lose sight of the fact that a lot of flying is done at and in the vicinity of airports where there is no tower and no one on the ground controlling traffic, whether in the pattern, landing, or taking off. It's at these airports where radio communication, while not literally mandated, is so important. Go back and look at the accidents cited in Chap. 2; most occurred at these uncontrolled fields, and most were the result of one form or another of pilot (not equipment) radio failure. Coupled with constant scanning of the skies for other traffic, the intelligent use of the radio in this uncontrolled airport environment is absolutely essential to the well-being of all who are operating in that environment.

That said, let's move now to a discussion of radio usage at these non-tower-controlled airports, beginning with the most basic of all—the "multicom" airport, which has no radio facility of any sort. If you're not familiar with that term, don't let it throw you. It'll come into focus in a minute.

4
Airspace Classifications: A Summary

HAVING REFERRED BRIEFLY IN PREVIOUS CHAPTERS TO "uncontrolled" airports, "controlled" airports, "controllers," "ATC," and the like, it's time now to become more specific as to the full meaning of these and similarly related terms. Particularly it is time because from here on I'll be mentioning the various classifications of airspace, and unless those references are understood, they'll only cause raised eyebrows and searches for definitions. What follows, then, are basically introductory descriptions of these airspaces and their principal operating requirements—sort of a foundation, if you will, because I'll have more to say later as we talk about radio procedures when flying into or out of both controlled and uncontrolled airspaces and their respective airports.

"CONTROLLED" AND "UNCONTROLLED" AIRSPACES DEFINED

By FAA definition, any airspace that is called "controlled" simply means that air traffic control service, including aircraft separation, is provided to IFR flights. For those not thoroughly familiar with the term, an IFR flight is one that has filed an instrument flight plan, is required to abide by all instrument flight rules, regardless of weather conditions, and is subject to air traffic control instructions. Air Traffic Control, or ATC, refers to FAA personnel in the control towers at airports, the Approach/Departure controllers at the major airports who oversee the arrival and departures of IFR and VFR aircraft, and the Air Route Traffic Control Center personnel who are responsible for controlling IFR flights as they move across the country from one airport to another.

While the emphasis here may seem to be on IFR operations, be aware that ATC must provide certain services to VFR aircraft and will provide others, such as advisories of conflicting traffic on cross-country flights, if the pilot has so requested and if ATC's workload permits. This will occur even if the pilot had filed no flight plan and was just out on a brief Sunday afternoon excursion.

"Uncontrolled" airspace is simply that in which ATC exercises no control over any type of flight—IFR or VFR. In a nutshell, that encompasses most of the airspace below 1,200 feet agl (above ground level) and all airports that have no operating control tower and no human or automated weather-reporting service (which is the vast majority of airports in the United States).

If you're a student or perhaps only a weekend pilot, don't let the word *controlled* mislead you. The fact that by definition most of the airspace is controlled in no way confines you to flights below 1,200 feet or only into nontowered airports— just the opposite. Instead, should you choose to do so, you could fly VFR from one end of the country to the other, always in controlled airspace, and never once talk to a single air traffic controller or Flight Service Station specialist. All you have to do is be aware of the various classes of airspace and the requirements to operate within them. Then, if you don't meet the pilot or operating requirements, or don't want to bother with them, just set out across the countryside, abide by the VFR flight altitude, visibility, ceiling, and cloud-separation regulations, and avoid those areas (mostly airports) that require contact with ground controllers.

Perhaps that observation leaves you scratching your head, but hang on. Hopefully I'll be able to clarify things as we move into later chapters. For now, though, let's focus on identifying the airspaces, along with very brief summaries of the operating requirements for each.

THE AIRSPACES DESCRIBED

One explanatory note first: The United States adopted the current airspace designations in September, 1993 in order to conform to those of the International Civil Aviation Organization (ICAO). Prior to 1993, we called today's Class B airspace a Terminal Control Area (TCA); an Airport Radar Service Area (ARSA) is now a Class C airspace; the Positive Control Area (PCA) is today's Class A airspace, and so on. What we had then versus what we have now is of no consequence, but I mention it so that if you hear pilots talking about TCAs, ARSAs, PCAs or the like, you can smile knowingly, aware that they may be just a few old-timers clinging to yesteryear.

But back to today and the airspaces—to be summarized in the following. If you'll follow along with Fig. 4-1, you may find the various descriptions a little less confusing.

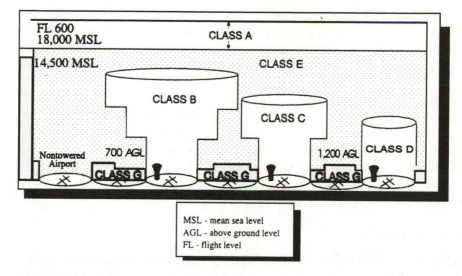

FL 600
18,000 MSL
CLASS A

14,500 MSL
CLASS E

CLASS B

CLASS C

Nontowered Airport
700 AGL
1,200 AGL
CLASS D

CLASS G
CLASS G
CLASS G

MSL - mean sea level
AGL - above ground level
FL - flight level

Fig. 4-1. *Think of the airspaces in terms of diminishing control: Class A—most controlled; Class B—next most controlled; Class C—next most, and so on to Class G— the least controlled.*

Class A Airspace

This is the airspace from 18,000 feet msl (mean sea level) up to 60,000 feet msl—or, as ATC would put it, "FL (Flight Level) 60." Here, the pilot-in-command must be instrument-qualified, all flights must operate in accordance with instrument flight rules (IFR), and all aircraft must have operating transponders with altitude-reporting capabilities. Other regulations do apply to flight in Class A airspace, but these are the basic requirements.

Class B Airspace

The Class B airspaces are those surrounding the nation's largest and busiest (currently 33) airports, based on the number of IFR operations or the volume of passenger enplanements. Dimensionally, the airspace extends vertically from the surface to approximately 8,000 feet msl and 25 to 30 nautical miles horizontally. As typically described, the Class B resembles an upside-down wedding cake, with a core center and then succeedingly larger layers extending out from the core. The radius of the core on the surface is approximately 5 nautical miles out from the primary airport. At the higher levels, layers, or shelves (the terms are interchangeable), the radius varies from 10 up to 30 miles, depending on the number of levels and the shape of the airspace at a given airport. Because of the heavy concentration of traffic in these B airspaces, penetration of any portion of the structure requires specific approval from the responsible controlling unit. At Class B and Class C terminals, that unit is called Approach/Departure Control.

To operate in a Class B airspace, pilots must hold at least a private pilot certificate (there are some exceptions for student pilots, which I'll summarize in Chap. 12).

Equipment-wise, for VFR operations, the aircraft must have a transponder with altitude-reporting capabilities (a Mode C transponder) and a two-way radio capable of communicating with ATC on the frequencies of the particular Class B airspace. IFR operations require the same radio and transponder plus an operable VOR or TACAN receiver.

A Class B airport is easily identified on sectional charts by a large blue square that encompasses the primary airport and its surrounding geography. This square, incidentally, reflects the territory covered by the Terminal Area Chart—a chart that enlarges the territory within the square and makes it easier for pilots not familiar with the area to spot landmarks and reporting points. The other identifying feature of a Class B airspace is the structure of the airspace itself and the heavy blue circles or designs that establish the altitudes and extensiveness of the levels that comprise the airspace.

Class C Airspace

Similar to the Class B, the Class C airspace refers primarily to the airport environment—in this case, some 120 airports with high levels of flight activity but not at the level of a Class B. Similar to the Class B, this airspace can be spotted on the sectional chart by the blue airport symbol and the magenta circles surrounding the primary airport. The typical Class C structure has a central core with a 5 nautical mile (NM) radius, one higher level with a 10 NM radius, and a common ceiling of about 4,000 feet agl. These are called *inner* and *outer circles.* Beyond the 10 NM radius of the outer circle, there's another 10 miles of airspace called the *outer area.* I won't confuse you with that feature now, but we'll come back to it in Chap. 11.

As to operating regulations: Contact with ATC is required before entering the area; any pilot, from student on up, may operate within the airspace; and the aircraft must have a two-way radio and an operable transponder with altitude-reporting equipment (Mode C).

Class D Airspace

The Class D airspace is designed for still smaller airports but those that have sufficient traffic volume to justify a control tower. In reality, some of these Class D airspaces can be busier in a month or a year than a Class B airport, but the traffic is principally of a light plane VFR nature—not IFR or commercial with high passenger enplanements. As a rule, the Class D airspace rises from the surface to about 2,500 feet agl, and a blue segmented circle on the sectional denotes the five nautical mile radius of the airspace around the airport.

Any pilot, student on up, can operate within the airspace and the aircraft need have only an operating two-way radio capable of communicating on the tower frequency. To identify a Class D airspace on the sectional chart, look for the blue airport symbol. All airports with towers are colored blue on the sectional charts, but Class B and C airspaces have the large circular designs around them, which, other than the segmented circle, is not true of the Class D fields.

Class E Airspace

As *AIM* says, "Generally, if the airspace is not Class A, B, C, or D, and it is controlled airspace, it is Class E airspace." Further, as a general rule, the Class E airspace rises from the surface, or from a designated altitude, up to 18,000 feet msl, which is the floor of the Class A airspace. And even further as a general rule, barring interruptions by Class B, C, or D airports and the Class G uncontrolled airspace, Class E is everything else. Or, putting it another way, and disregarding Class A for now, think of the entire airspace as Class E, broken up only by the Class B, C, and D airports and the uncontrolled Class G. This means that all of the airspace, except for Class G, is controlled—which means, again, that ATC provides control to all IFR operations from Class A through Class E.

And that includes Class E airports as well. Check a sectional chart and you'll find a lot of airports depicted in magenta, most of which are non-tower-controlled Class G airspaces. (*Note*: Formerly referred to as "uncontrolled" airports, the FAA terminology is now "non-tower-controlled.") Some, however, are surrounded by a single segmented magenta circle and probably, but not necessarily always, inclusion of "ASOS" or "AWOS" in the airport identification area. These are Class E airports which have either a human or automated weather reporting source (ASOS or AWOS) on the airport property and are thus controlled airspaces for IFR operations in IMC (instrument meteorological conditions) weather. If there is no weather-reporting source, ATC will not exercise control of IFR aircraft for landing or takeoff operations, and the airport is thus a Class G. Even a Class D reverts to a Class G airport if the D tower closes down at night and there is no weather information source on the airport.

Class G Airspace

No, I didn't leave anything out: There is a Class F airspace, but that's for foreign use only. We in the United States have nothing to correspond with it, so forget about it, unless you intend to go flying overseas.

Class G is what little airspace Classes A, B, C. D, and E leave untouched—meaning the space from the surface to either 700 feet agl in IMC around some airports or 1,200 feet agl—the Class E floor. Pilot minima are student only and the aircraft need have no radios.

This is obviously the simplest environment in which to fly, from a traffic control point of view. Keep in mind, though, that you have only a maximum altitude of 1,200 feet to play with, so except for pattern work around a non-tower-controlled airport, the Class G operating space to go places and do things is somewhat limited.

Terminal Radar Service Areas (TRSAs)

One other airspace needs to be mentioned—the Terminal Radar Service Area, or more commonly referred to as TRSAs. Identified on sectional charts by black lines (they appear as almost a dark gray) that surround a tower airport, the TRSA is basically circular in structure and extends outward about 15 nautical miles in a pattern somewhat similar to the Class B and C airspaces. The airport itself is a Class D, but the rest of the airspace underlying the TRSA is normally Class E.

What makes the TRSA unique is that the tower, as usual, is responsible for all traffic within the approximate 5-mile radius of the airport, but while Approach/Departure Control provides radar service to all IFR aircraft beyond that 5-mile radius, elements of the service are *available* to arriving and departing VFR traffic.

That needs explaining. Note the word *available.* It simply means that arriving VFR pilots can enter the TRSA area (not the Class D airspace) without obtaining permission from Approach Control. They simply advise approach of their N-number, position, attitude, and intentions. Then, all that those who don't want the radar service have to do is conclude this introductory call with "Negative TRSA service."

On the other hand, if they do want the service, they end the introductory call with "Request TRSA service." Approach will then provide traffic advisories of other aircraft in the airspace, limited vectoring, and, at locations where the procedures have been established, sequencing of aircraft before turning the aircraft over to the control tower for traffic pattern and landing instructions.

The reason TRSAs still exist as unlettered airspaces is that they lie somewhere between Class C and D airports in terms of IFR activity and enplanements. Also, ICAO has no comparable airspace—hence no matching alphabetizing that we could use as identification. There aren't too many of these TRSAs around the country, and sooner or later, it's expected that each will move up to a Class C or down to a Class D airspace. Meanwhile, they're sort of an anomaly and probably won't remain very long in that "neither-nor" state as part of the airspace system.

SPECIAL USE AIRSPACE (SUA)

There is one more category of airspace that must be mentioned—the Special Use Airspace (SUA). In a word, SUAs are blocks of space established for purposes of national security, welfare, or environmental protection, as well as for military, research, testing, development, and evaluation (RTD&E).

In summary, the types of SUAs are as follows.

- *Prohibited:* These are specifically defined areas on the surface of the earth over which flight below a certain published altitude is prohibited. Examples of such are the nation's capitol, Camp David, presidential residences or retreats, nuclear testing areas, and similar critical military or government facilities.

  Prohibited areas are relatively few in number and small in area, and are thus not major obstacles to a straight-line flight. Identified on the sectional chart by a blue design of small inward-pointing lines and on the legend flap, pilots should be aware of the dimensions of these strictly prohibited areas and take whatever action is necessary to avoid them. Prohibited means *prohibited.*

- *Restricted:* Larger and more common are these restricted areas in which there may be artillery fire, aerial gunnery, guided missiles, and similar military activities. Penetration of a restricted area without approval from the using or controlling authority is not only illegal but could result in a rather exciting and hazardous adventure. There should never be a reason for accidental penetration, though, because the areas are easy to spot on the sectional chart with their blue design and small inward-pointing lines, similar to the prohibited area identifier. They are also listed on the back of the sectional chart's legend flap, along with the altitudes within which they are active and the times of use.

If the using agency (as the Army or Air Force), because of temporary non-use, releases the airspace back to the controlling agency [usually the Air Route Traffic Control Center (ARTCC) in which the area is located], you can enter the restricted airspace *if* you have contacted that ARTCC and entry has been approved. Otherwise, don't go by "Time of Use" on the sectional chart's legend flap. Things could, and do, change. Check with Center first.

• *Warning areas:* These areas are located offshore, beginning 3 miles out. Accordingly, they should be of little concern to the average VFR pilot. Just be aware that there are such areas that could contain military activity potentially hazardous to nonparticipating aircraft.

• *Military Operations Areas* (*MOAs*): Here are the largest chunks of airspace that could get in your way on a cross-country VFR flight. Identified on the sectional chart by small magenta inward-facing lines, MOAs are reserved for military flight training, which includes all sorts of aerobatics, unusual positions, and maneuverings that demand the full attention of the pilot. While VFR traffic is not prohibited from entering an active MOA ("hot," as the expression goes), it would be foolhardy to do so without first calling the responsible ARTCC to determine the true status of activity in the area. If it is hot, go around the whole airspace. Those Air Force pilots won't be looking out for civilian traffic in their midst, and the aircraft they're flying, as you know, could be on you before you had a prayer of taking evasive action. But, should the ARTCC tell you that all is clear, proceed ahead as though the MOA never existed.

• *Military Training Routes* (*MTRs*): Watch out for these! They're just thin gray lines on the sectional chart, but they identify the IFR (labeled "IR") and VFR (labeled "VR") routes used for, yes, military flight training. And, these routes, innocuous though they seem on paper, are much wider than the lines would imply and wider than the standard Victor airways. The latter are 4/4, meaning 4 miles wide either side of the center line, but some of the MTRs will run 5/5, 10/10, or even, as one does, 16/25.

If your route parallels or crosses an MTR, check with the appropriate Flight Service Station to get a reading of the potential activity at the time you will be in the MTR vicinity. The FSS will have the intended activity, but once on your way you'd be wise to call the controlling Center and ask for an update. Military plans and schedules being subject to change, the Center will always have the latest schedule of activity.

• *Alert areas:* One other airspace area is the alert area. Again identified on the sectional by the blue inward-pointing lines, this is not a prohibited area in any way. It exists and is charted simply to warn all pilots of an ongoing high level of flight activity in the area and thus the need for special alertness to what's going on outside the cockpit window.

A quick review of the Special Use Airspaces follows. If it's not the case already, you should be aware of what these airspaces are and their potential effect on VFR flight. In essence, stay out of prohibited, restricted, and warning areas. Enter MOAs only after determining that they're not active. Determine the level, type, and times of traffic on the MTRs, and then keep your eyes open when you're on or crossing one of the routes. Finally, be alert for other traffic while you're in an alert area. SUAs of whatever type are really not very good places in which to play around!

CONCLUSION

As this is merely an overview of the airspace system, I've intentionally omitted excerpts from sectional charts and the *Airport/Facility Directory* that would illustrate the various airspaces more clearly than just the written word. Those excerpts will come later, though,

Airspace	Flight Visibility	Distance from Clouds
Class A	Not Applicable	Not Applicable
Class B	3 statute miles	Clear of Clouds
Class C	3 statute miles	500 feet below 1,000 feet above 2,000 feet horizontal
Class D	3 statute miles	500 feet below 1,000 feet above 2,000 feet horizontal
Class E Less than 10,000 feet MSL	3 statute miles	500 feet below 1,000 feet above 2,000 feet horizontal
At or above 10,000 feet MSL	5 statute miles	1,000 feet below 1,000 feet above 1 statute mile horizontal
Class G 1,200 feet or less above the surface (regardless of MSL altitude).		
Day, except as provided in section 91.155(b)	1 statute mile	Clear of clouds
Night, except as provided in section 91.155(b)	3 statute miles	500 feet below 1,000 feet above 2,000 feet horizontal
More than 1,200 feet above the surface but less than 10,000 feet MSL.		
Day	1 statute mile	500 feet below 1,000 feet above 2,000 feet horizontal
Night	3 statute miles	500 feet below 1,000 feet above 2,000 feet horizontal
More than 1,200 feet above the surface and at or above 10,000 feet MSL.	5 statute miles	1,000 feet below 1,000 feet above 1 statute mile horizontal

Fig. 4-2. A means to check at a glance the VFR visibility and ceiling separation minima within the various airspaces.

Airspace features	Class A	Class B	Class C	Class D	Class E	Class G
Operations permitted	IFR	IFR and VFR	IFR and VFR	IFR and VFR	IFR and VFR	IFR and VFR
Entry requirements	ATC clearance	ATC clearance	ATC clearance for IFR. All require radio contact.	ATC clearance for IFR. All require radio contact.	ATC clearance for IFR. All IFR require radio contact.	None
Minimum pilot qualifications	Instrument rating	Private or student certificate	Student certificate	Student certificate	Student certificate	Student certificate
Two-way radio communications	Yes	Yes	Yes	Yes	Yes for IFR	No
VFR minimum visibility	N/A	3 statute miles	3 statute miles	3 statute miles	3 statute miles[1]	1 statute mile[2]
VFR minimum distance from clouds	N/A	Clear of clouds	500' below, 1,000' above, and 2,000' horizontal	500' below, 1,000' above, and 2,000' horizontal	500' below, 1,000' above and 2,000' horizontal[1]	Clear of clouds
Aircraft separation	All	All	IFR, SVFR, and runway operations	IFR, SVFR, and runway operations	IFR and SVFR	None
Conflict resolution	N/A	N/A	Between IFR and VFR ops	No	No	No
Traffic advisories	N/A	N/A	Yes	Workload permitting	Workload permitting	Workload permitting
Safety advisories	Yes	Yes	Yes	Yes	Yes	Yes
Differs from ICAO	No	Yes[3]	Yes[3,4]	Yes for VFR[4]	No	Yes for VFR[5]

[1] Different visibility minima and distance from cloud requirements exist for operations above 10,000 feet msl

[2] Different visibility minima and distance from cloud requirements exist for night operations above 10,000 feet msl, and operations below 1,200 feet agl

[3] ICAO does not have speed restrictions in this class—U.S. will retain the 250 KIAS rule

[4] ICAO requires an ATC clearance for VFR

[5] ICAO requires 3 statute miles visibility

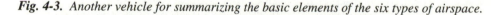

Fig. 4-3. *Another vehicle for summarizing the basic elements of the six types of airspace.*

in the chapters that focus on the various controlled and non-tower-controlled airports and operations in the Class E spaces.

For a few more details about these airspaces right now, though, you might want to study Figs. 4-2 and 4-3. Fortunately, the whole airspace structure, and indeed the system itself, is relatively uncomplicated—even for those new to aviation. But if one is to use it and fly in it with safety, it's a system, along with its various operating rules and requirements that must be understood. Hopefully we've taken some steps here to enhance that understanding, so let's move on now to the most basic environment of all in which to fly: a small, no-tower airport, typically just a grass or asphalt strip in a Class G airspace—the airport that is commonly called a *multicom* airport.

5
Multicom Airport Radio Communications

A FEW YEARS AGO, I WAS DRIVING WITH A FRIEND DOWN A WELL-traveled street in Riyadh, Saudi Arabia. Many side roads intersected this main street, but walls or buildings blocked the driver's view, making it impossible to see any cross traffic that might be approaching. Now driving in Saudi Arabia is a thrill in itself, but when there are few stop signs or traffic lights and you can't see cars that might jump out at you from the left or the right, extreme caution is the only alternative for your continued physical well-being.

In this case, my driver friend slowed down at every blind intersection and honked his horn. At night, he honked the horn while blinking his lights. These signals alerted others that he was there. They were his nonverbal communication signifying his presence as well as his intentions.

In a more sophisticated sense, multicom is akin to the horn and lights of our Saudi driver. At Class G airports which have no ground-based traffic control or advisory service, no "red or green lights," no "stop signs," multicom provides the aural communications that reveal your presence and your intentions. Just like the Saudi driver, you are transmitting in the blind to anyone who is tuned to the standard multicom frequency.

You don't know if anyone is really listening or is even in the immediate vicinity, but like the driver, you take that added step—just in case.

In its simplest terms, multicom is nothing more than communication between two aircraft, whether in the traffic pattern or flying at altitude along the same route. While it can be comforting and perhaps important to talk to another pilot during a cross-country flight to exchange weather information and informal pilot reports (PIREPs), the real value of multicom comes to the fore around uncontrolled, nonunicom airports. That's when you need to know who is there, where they are, and what they intend to do—just as they need to know the same about you. Multicom provides the vehicle for that exchange of information over a common frequency.

The key here is a common frequency. The FAA has thus established what it calls Common Traffic Advisory Frequencies (CTAFs). The CTAFs may be for airports that have no ground radio facilities at all, and are thus referred to as "multicom" airports, those that have only field-advisory radio services provided by the local Fixed Base Operator (unicom airports), Flight Service Stations, or tower-controlled airports that operate only part time.

Two sources tell you what the CTAF is for a given airport. One is the *Airport/Facility Directory (A/FD)*, as illustrated in Fig. 5-1. The other is the sectional chart (Fig. 5-2). If you refer to the sectional chart, you'll note the small circle with the letter "C," which indicates the CTAF for that airport. Immediately to the left of the circle is the frequency itself, in this case, 122.9. CTAFs that are printed in slanted, or italic, numbers and are magenta in color identify non-tower-controlled airports.

Keep in mind these points about CTAFS: (1) "Common" doesn't mean one universal frequency for all airports but rather the common frequency that all pilots should use for a given airport; (2) Despite what I just said, multicom—and only multicom—airports do have one common frequency nationwide—and that frequency, again, is 122.9. Whatever the case, however, reference to the sectional chart or the *A/FD* is essential to determine the type of airport you are planning to enter and its correct radio frequency.

But back to operating in a multicom airport environment. Despite the importance of radio communications in such an environment, let's not be naive. No matter how clearly and explicitly you transmit your intentions, not all aircraft have radios. Even

---

WAMEGO MUNI (69K)   3 E   UTC–6(–5DT)   N39°11.83' W96°15.53'   **KANSAS CITY**
L–6H
966   B   FUEL 100LL
RWY 17–35: H3170X30 (ASPH)   RWY LGTS (NSTD)
    RWY 17: Thld dsplcd 170'. P-line.   RWY 35: Trees.
AIRPORT REMARKS: Unattended. Parachute Jumping. Ultralight activity on and in vicinty of arpt. 8' ditch across apch
    end Rwy 17. For fuel call Wamego Police Dept 913–456–9553. Rwy 17–35 NSTD LIRL. lgts placed 25' from rwy
    edge. ACTIVATE LIRL Rwy 17–35—122.9.
COMMUNICATIONS: CTAF 122.9 ◄—
    WICHITA FSS (ICT) TF 1–800–WX–BRIEF. NOTAM FILE ICT.
RADIO AIDS TO NAVIGATION: NOTAM FILE MHK.
    MANHATTAN (T) VORW/DME 110.2   MHK   Chan 39   N39°08.73' W96°40.12'   075° 19.4 NM to fld. 1060/6E.
    HIWAS.

Fig. 5-1. The Airport/Facility Directory (A/FD) identifies the Wamego CTAF as 122.9.

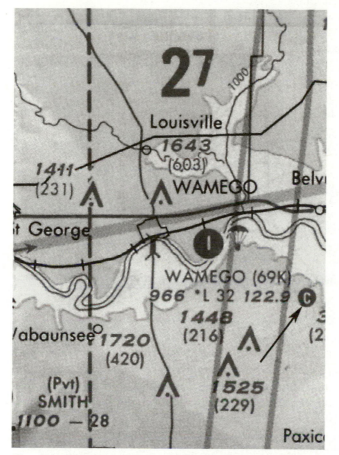

Fig. 5-2. *The sectional chart also identifies the Wamego airport and its 122.9 multicom frequency.*

if they do, their pilots may not be tuned to the 122.9 frequency. Indeed, they might not have their radios on at all. ("Why bother about such things at Peapatch Municipal?")

The alternative is obvious. Flying around an uncontrolled airport demands a swivel neck and sharp eyes. To rely solely on blind transmissions is to invite a few thrills or unexpected encounters of the worst kind. Keep in mind, too, that your transmitter might suddenly go on vacation. You think you're broadcasting to all and sundry, but nothing is passing beyond the mouthpiece of your mike. Such failures have happened—and they could happen to you anytime, anywhere. Open eyes and constant head-turning are your best defenses against near hits or close misses.

WHY USE MULTICOM?

Why use multicom? The answer is already evident: safety. If you use it, other aircraft in the area might hear you and use it too. Then everyone will be informed about who is doing what, where, and when.

At the same time, don't be dumb. Be willing to back off when judgment so dictates. You've done a good job of advising others of your actions and intentions, but just as you're about to turn on final approach, you see some guy coming from nowhere on a long straight-in approach. Decision time—do you assume he knows you're there? Do you assume he'll give way because you're apparently number one to land? Go ahead and assume, but it's a dangerous practice.

Discretion says give way. In this case, it's better to be number two than arrive at the same point in a dead heat. Piggybacking might be the way to get a space shuttle to Cape Kennedy, but it's not a very comfortable way for a Cherokee and a Cessna to make a landing. That's a "dead heat" in a very literal sense.

A patently obvious observation? If it is so obvious, think about the examples in Chap. 2. Why do we still have midair collisions, landing accidents, and an excessive number of close calls? Part of the reason might be complacency ("It can't happen to me.") Part could be ignorance or stupidity—or a combination of the two. In some cases, it's nothing but a flagrant disregard for the rights of others. Whatever the reason, the diligent use of multicom at least reduces the potential of trouble while helping others who might be as concerned about their well-being as you are about yours.

USING MULTICOM

This is repetitous, but it should be said again: as with any transmission, follow the Navy's admonition about report writing and correspondence: KISS—Keep It Simple, Stupid. Know what you're going to say, say it in plain English, say it clearly, and keep it short. If you've practiced your radio technique as suggested, you know how you sound. You should have learned to speak distinctly and slowly enough to be understood and to convey your message in an organized sequence. If you speak with the listener in mind, you'll communicate more effectively with fewer words and less monopolizing of the airwaves.

A SIMULATED LANDING AND DEPARTURE WITH MULTICOM

With the "theory" out of the way, let's go through the multicom procedures when operating into and out of an uncontrolled airport. Some of the transmissions might also seem repetitious, but keep in mind that safety is the first concern. A single transmission might be the one that saves the ship.

Approach to the Field

Tune to 122.9 at least 10 miles out. If there's any activity, you might get an idea of the traffic volume and pick up the favored runway and wind direction. Assuming this to be the case, start the self-announcing process by identifying your aircraft, position, altitude, and intentions. As there is no control agency of any sort on the field, your call is designed to advise other aircraft of your presence in the area. Consequently, open the transmission with the name of the airport, followed by "Traffic." Let's assume that you're going into

Wamego (Kansas) Municipal Airport (Figs. 5-1 and 5-2), with its elevation of 966 feet and a 17–35 runway. Let's further assume that you've heard at least one other aircraft report its position in the traffic pattern and have learned that Runway 17 is the runway in use. The call, then, would go like this:

> Wamego Traffic, Cherokee One Four Six One Tango is ten north at five thousand five hundred. Will enter left downwind for full stop One Seven Wamego.

Note that the name of the airport is repeated at the end of the transmission. This is to make certain that there is no confusion on the part of other aircraft as to the airport to which you are going or at which you are operating. This positive identification is especially important when two or three other airports are in the general vicinity, each with several aircraft in the air, and all transmitting on 122.9. Identifying the intended airport twice minimizes the potential for confusion.

Going back to the arrival example, if you hear nothing after tuning to 122.9, don't be lulled into the belief that the skies are clear. Give yourself every safety edge you can. Announce your intentions to those who might be listening:

> Wamego Traffic, Cherokee One Four Six One Tango is ten north at five thousand five hundred. Will cross midfield at two thousand five hundred for wind tee check and landing Wamego.

Over the Field

Assuming that you've heard no other traffic, you're over the field and see that the wind tee or sock favors Runway 17. Again, announce your position and intentions to the seen or unseen audience:

> Wamego Traffic, Cherokee Six One Tango over the field at two thousand five hundred. Will enter left downwind for Runway One Seven, full stop, Wamego.

Although the following was mentioned back in Chap. 1, perhaps a bit more explanation at this point is in order: Note the omission of two digits when stating the aircraft's N-number in this second call: Instead of "Cherokee One Four Six One Tango" the identification has been shortened to "Cherokee Six One Tango." After initially transmitting the full identification, succeeding calls can be reduced to aircraft type plus the last two numbers and the assigned phonetic alphabet letter (or the third number, if the aircraft has no assigned letter, as is sometimes the case).

The only time this call-sign shortening should be avoided is when there could be confusion with another plane in the area or pattern which has a somewhat similar call sign. For example, using the abbreviated "Cherokee Six One Tango" as the base for comparison purposes, you might find closely related shortened alphanumerics such as "Cessna Six One Tango," "Cherokee Six TwoTango," "Mooney Six One Papa," or another Cherokee with the five-character call sign of "Cherokee Eight Three Six One Tango." In that admittedly rare coincidence, it would be just a little confusing to other aircraft in the vicinity to have two "Six One Tangos," both Cherokees, flying around in the same general area. Once duplication of the three-character designation is discovered, both aircraft should revert to their full character.

Barring such N-number similarities or sources of possible confusion with another aircraft, this sort of verbal shorthand, after your initial full identification and position call, is very much in order. It shortens succeeding calls while conserving air time. Those within the sound of your voice will appreciate your thoughtfulness.

Entry to Downwind

You're entering the downwind leg at pattern altitude—approximately 800 feet agl. Get on the air again:

> Wamego Traffic, Cherokee Six One Tango entering left downwind for Runway One Seven, full stop, Wamego.

Turning Base

Going into the turn from downwind, make your next call:

> Wamego Traffic, Cherokee Six One Tango turning left base for Runway One Seven, full stop, Wamego.

At any uncontrolled airport, make this call when turning *onto* the base leg. It's a lot easier for other aircraft to see you when you're in a bank as opposed to straight-and-level flight. Also, a fairly wide pattern with a distinct base leg is preferable to a hotshot U-turn. This gives you time to scan the area for other aircraft on the same leg that haven't announced their presence. It also allows you to check the final approach course for someone who might be making a straight-in approach. This is the altar on which many inflight marriages have been consummated—unwanted but nevertheless eternal marriages.

Turning Final

Once again, announce what you're doing while in the turn to final approach:

> Wamego Traffic, Cherokee Six One Tango turning final for Runway One Seven, full stop, Wamego.

Clear of the Active

On the ground, get off the runway as quickly but safely as possible. Somebody you never heard of might be on your tail. When clear, don't keep it a secret:

> Wamego Traffic, Cherokee Six One Tango clear of One Seven, Wamego.

A lot of talk? Yes, but at least you've fulfilled your responsibilities. You've kept others informed, and you've made the air just that much safer. Now if only somebody as considerate is listening....

You've gassed up, coffeed up, and are ready to go again. Once the engine has fired and the radio is on, listen for a moment or two while you're still on the ramp to see if any traffic has developed. As in the prelanding, your actions can be planned according to what you hear—or don't hear. Regardless, don't assume! Announce your intentions. At most one-strip fields a taxiway is an unknown luxury. Back-taxiing on the active runway is the only way to get into takeoff position. But let the other guy, if he's out there, know what you're going to do.

Taxi and Back-Taxi

Assuming that there is no taxi strip and that you'll have to taxi back (or "back-taxi") on the active runway for takeoff, don't make your initial call when you're still on the ramp, which may be some distance from the runway itself. Instead, move out to the hold line and set the brakes. Then, if there's a runup area at the departure end of the runway, proceed with your call thusly:

> Wamego traffic, Cherokee One Four Six One Tango, back-taxiing Runway One Seven for takeoff Wamego.

On the other hand, if there is no runup area at the end of the runway, taxi to the hold line and go through the complete pre-takeoff checklist and engine runup. When you're finally ready to go, make this initial call:

> Wamego traffic, Cherokee One Four Six One Tango back-taxiing Runway One Seven for takeoff, departing to the (direction).

Or, if you're staying in the pattern for touch-and-gos:

> Wamego traffic, Cherokee One Four Six One Tango back-taxiing Runway One Seven, closed pattern, Wamego [or "closed traffic," or "for touch-and-gos"; "closed pattern" is the correct phraseology, however].

If you do have to back-taxi on the active runway, first scan the approaches for both Runways 17 and 35 before venturing on to the runway itself. It's possible that some unheard-from individual is landing downwind or is indeed on the final approach for Runway 17 but hasn't bothered to tell anybody. But, if the air is clear, apply power and get to the end as quickly and safely as possible. Otherwise, an approaching aircraft might have to go around. Or worse yet, its pilot might not see you in time to abort the landing—and that could develop into a rather messy situation. Then, if you've completed the pre-takeoff check, do a 180-degree turn, apply the power, and get going. This is no place to dillydally.

Preflight Runup

Assuming, on the other hand, that there is a runup area, and if that area is so structured, back-taxi to the end of the runway and then turn left so that you can park on the departure runway's *right* side. The reason for this positioning is that after you've completed the engine runup, magneto checks, and all, and are ready to go, you can turn the plane so that it is at a 90-degree angle to the departure runway. Now, sitting in the left seat, you'll have a clear view of the final approach and any unannounced aircraft that might be about ready to land. If you were on the other side of the runway, and in the left seat, scanning the final approach area would be much more difficult, particularly in a high wing plane.

This might be a small point, but I've had more than one pilot pull out on the runway when I was on final approach. They simply hadn't seen me, and a go-around was the only alternative. Were they using multicom? That's a silly question.

Taking the Active Runway

While still in this 90-degree position, look upwind as well as downwind (some pilots, at this point, even swing a complete 360 degrees so they can scan the whole traffic pattern). Then, when you're sure everything is clear, make your departure call:

> Wamego traffic, Cherokee One Four Six One Tango taking Runway One Seven, departing to the east. Wamego.

On the matter of someone landing downwind, four conditions could bring about such a landing: (1) very light wind; (2) the pilot's complete disregard of the wind tee or sock; (3) the pilot's failure to monitor or use multicom; or (4) a genuine emergency. With observant eyes and the radio tuned to 122.9, there is no excuse for aircraft landing in opposite directions at an uncontrolled airport. And yet, because of pilot ignorance, lack of radio equipment, or failure to use the equipment, this sort of thing happens too frequently.

To wit: I recently saw an individual barrel-in downwind, narrowly missing a landing plane that was doing everything correctly. He discharged his passenger and then roared off from the taxiway-runway intersection, which left him about 2,500 feet of a 4,000-foot strip. This time, however, he went into the wind. Did this hotshot in his sleek Bonanza use the sophisticated radio equipment that the plethora of antennas implied? Not once. He was apparently above such trivialities.

Departure

Although you've already stated your flight intentions, it doesn't hurt to repeat them, once airborne, as a courtesy to those still in the pattern:

> Wamego Traffic, Cherokee Six One Tango off Runway One Seven, departing the pattern to the east, Wamego.

Until you're about 10 miles out, stay tuned to 122.9 to pick up any traffic that might be inbound, in your line of flight, or at your altitude. Also, you could help an arriving pilot who has just made an initial call by giving him or her the wind and runway information. For example:

> Aircraft calling Wamego Traffic, Cherokee Six One Tango just departed Wamego. Favored runway is One Seven, winds about two two zero at six.

There's no set pattern for such a call, so help the other pilot in your own words. Use the word "favored," however, when giving runway information. "Active" implies that a particular runway must be used—which, at an uncontrolled airport, is not the case.

Touch-and-Gos

Let's suppose that instead of departing the pattern, you want to make a few touch-and-gos. If you're parked at the ramp, the calls are the same up to the time you're ready to take off. Then get on the air:

> Wamego Traffic, Cherokee One Four Six One Tango taking Runway One Seven, closed pattern for touch-and-gos, Wamego.

On the downwind, turning base, and turning final approach, repeat your intentions, just as you did for the full-stop landing:

> Wamego Traffic, Cherokee Six One Tango turning downwind for Runway One Seven, touch-and-go, Wamego.
> Wamego Traffic, Cherokee Six One Tango turning base for Runway One Seven, touch-and-go, Wamego.
> Wamego Traffic, Cherokee Six One Tango turning final for Runway One Seven, touch-and-go, Wamego.

Yes, that's a total of four messages for one touch-and-go, but you never know who has just entered the pattern. Your last message could be the first he has received. You can't be sure—so be safe.

When you've had enough for the day, you're going to land or leave the pattern. In either case, keep the other traffic informed. If it's the final landing, make the downwind, base, and final calls similar to those cited above, substituting "full stop" for "touch-and-go":

> Wamego Traffic, Cherokee Six One Tango turning downwind for Runway One Seven, full stop, Wamego.

If you're leaving the pattern, make the call after the last takeoff and when you have the aircraft safely under control:

> Wamego Traffic, Cherokee Six One Tango departing the pattern to the east, Wamego.

CONCLUSION

Inexperienced pilots; pilots who don't know how to maneuver in even moderate traffic; failure to give way to others; lack of knowledge of multicom uses and techniques; failure to turn on the radio; no radio at all: all are reasons for accidents at uncontrolled airports.

In many respects, the controlled airport, even with its five o'clock congestion, creates a greater feeling of security than setting down at a small-town, uncontrolled field. The forced and enforced radio communications make the difference. At hundreds of fields like Wamego, vigilance coupled with skillful use of multicom greatly enhances the safety and mental tranquility all pilots seek in flight.

6

Unicom Airport Radio Communications

THE MOST MODEST AIR-TO-GROUND COMMUNICATION (AND ALSO one that can be very helpful) is provided by unicom. In a sense, it's a step up from multicom and a step below the tower communications in a controlled airport traffic area.

WHAT IS UNICOM?

Simply stated, unicom permits radio contact with a ground facility on the airport. At many locations without a tower or Flight Service Station, the unicom operator fills the void by giving "field advisories" to pilots who call in and request them. The advisory consists of wind direction and velocity, possibly altimeter setting, the favored runway, and any reported traffic. Unlike a tower, however, unicom is not a controlling agency. The operator gives information, and that's all. The rest is up to the pilot.

Unicom also provides services of a nonflight nature. For example, if you want a fuel truck available for a quick turnaround, the unicom operator can make the arrangements. Maybe you need a taxi, or you'd like the operator to call your office or home to advise someone of your arrival time. Unicom is there to help you.

Of course, if an operating tower or FSS is on the field, unicom can't and won't give you runway, winds, or traffic information. That's the responsibility of the official facility. Unicom will, however, provide the nonflight services mentioned.

WHO OPERATES UNICOM?

At uncontrolled airports (the primary concern at the moment), the fixed-base operator (FBO) usually mans the unicom. The radio facility itself can be located anywhere at the airport, but it's generally in the lounge area where a call-in can be handled by the FBO manager or a jack-of-all-trades employee who answers the phone, keeps the books, and sells candy and sectional charts. Typically, a barometer and wind speed/direction indicator are near the transceiver.

Keep in mind that the unicom operator is not a controller. Indeed, he might have only the most meager knowledge of what goes on in the air. He probably will give you the best information he has, but it's unwise to count on 100-percent reliability.

For example, the wind direction, its velocity, and the favored runway might be completely accurate. The operator then concludes with "no reported traffic." In reality, half a dozen planes could be in the pattern, but none has been using the unicom frequency for that particular airport. Admittedly, if six airplanes are flying around him, it's a bit far-fetched to believe that he doesn't know that traffic is in the area; the point, however, is that there has been no radio contact with him or on his frequency. In effect, there's "no reported traffic."

Don't disbelieve the unicom operator, but learn not to depend on his every word. He simply might not be in a position to know everything that's going on outside. Perhaps he's not even a pilot. Or, as in every walk of life, he could be one of the few who just doesn't care. The moral is to use the service, but be vigilant as you enter the airport area. Your own eyes are the best instruments you have to tell you what's happening in the real world.

HOW DO YOU KNOW IF AN AIRPORT HAS UNICOM?

The easiest way is to check the sectional chart. Just to be sure there is no confusion about what that chart tells you in that regard, let's look at a couple of examples.

Figure 6-1 shows an uncontrolled airport (so identified on the chart by the magenta coloring of the airport symbol and the related data). Starting at the twelve o'clock position, the airport is Fine Memorial in Missouri's Lake of the Ozarks and is equipped with AWOS-3, an Automated Weather Observing System, available on the 135.325 frequency. Elevation of the airport is 869 feet above sea level, the runway is 6500 feet long, and the CTAF unicom frequency is 122.8.

The white diagonal line in the airport symbol identifies the one runway and its general northeast-southwest direction. Specifically, the headings are 30 and 210 degrees, but that information is obtained by reference to the government-published *Airport/Facility Directory*, or *A/FD*. The series of small dots surrounding the airport symbol means that a nondirectional beacon (NDB) is located on the field proper. The rectangular box

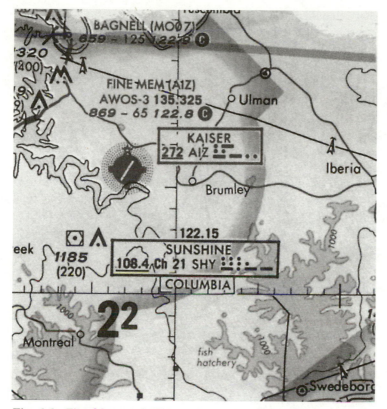

Fig. 6-1. *Fine Memorial (Kaiser) is a typical unicom airport, as depicted on the sectional chart.*

further identifies the NDB as "Kaiser" and, assuming the aircraft is equipped with Automatic Direction Finding equipment (ADF), the NDB can be accessed on frequency 272. Assurance that you have tuned in the correct NDB is determined by listening to the constantly repeated "AIZ" Morse code signal.

Finally, the star at the top of the airport symbol means that a rotating airport beacon is in operation from sunset to sunrise. The little ticks at the three, six, and nine o'clock positions signify that services are provided during normal working hours.

The sectional chart depiction, plus some additional symbology information on the chart's legend flap, provides a fair amount of information about a given airport, but it can't tell the whole story. The real details are best obtained from the appropriate *A/FD* (seven regional directories cover the continental United States, including Puerto Rico and the Virgin Islands, and all directories are published every eight weeks). Suffice it to say that a current sectional chart and *A/FD* are essential information sources and references if you're planning a cross-country flight. The only caution: If you have a sectional chart in your airplane, even if you're just shooting touch-and-go at your home airport, be sure that it's a *current* chart. Should an FAA inspector just happen to pull a ramp check and find an out-of-date chart somewhere in the cockpit, you, the pilot, will be cited for a Federal Aviation Regulation (FAR) violation. It's okay to have *no* sectional chart aboard, but it's not okay to have one that has expired. So be warned.

Going a step further, but still on the subject of unicom, Fig. 6-2 illustrates how the sectional chart depicts the Columbia Regional Airport, which is tower-controlled. (All symbols and data relating to such an airport are always colored blue on the sectional chart.) With or without coloring, the fact that a tower is on the field is established by the "CT" (Control Tower) and the tower frequency, in this case, "119.3." Next comes the small star reflecting that the tower operation is part time, followed by the "C," or CTAF symbol. That symbol means that transmissions should be made on the tower frequency of 119.3 even when the tower is closed. As to the other data, don't worry now about the "ATIS" reference or the "FSS" at the top of the data block (signifying the existence of a Flight Service Station on the field). Those will be discussed at the appropriate time.

Unicom service is also available at this airport, but the only indication of that is the italicized 122.95 frequency. In this instance, unicom could be contacted for whatever nonoperational services you might require, such as the need for a mechanic, a taxi, a call to the office, or the like. Otherwise, there's no reason to contact unicom because all flight-related information or instructions come from the tower controllers. When the

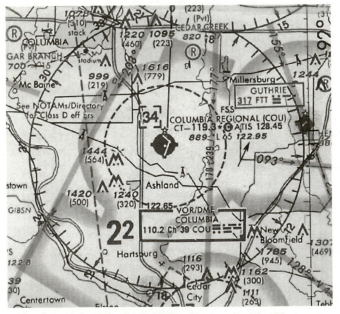

Fig. 6-2. Columbia Regional is an example of a tower-controlled airport with a unicom facility for nonoperational services.

```
COLUMBIA REGIONAL   (COU)   10 SE   UTC-6(-5DT)   N38°49.09' W92°13.18'              KANSAS CITY
  889   B   S4   FUEL 100LL. JET A   OX 2   ARFF Index A                              H-4G, L-21A
RWY 02-20: H6500X150 (CONC-GRVD)   S-92, D-125, DT-215   HIRL                             IAP
    RWY 02: MALSR. Tree.       RWY 20: ODALS. VASI(V4L)—GA 3.0° TCH 39'.
RWY 13-31: H4401X75 (ASPH)   S-24, D-24   MIRL
    RWY 13: REIL. VASI(V2L)–GA 3.0° TCH 44'. Road.    RWY 31: REIL. VASI(V2L)–GA 3.15° TCH 33'.
AIRPORT REMARKS: Attended Mon-Fri 1200-0600Z‡ and Sat-Sun 1300-0500Z‡. PPR for unscheduled air carrier
    operations with more than 20 passenger seats, call safety officer 314-443-2811. ARFF Index B level svc avbl
    on req. When twr closed ACTIVATE HIRL Rwy 02-20 MIRL Rwy 13-31, MALSR Rwy 02 and ODALS Rwy
    20—119.3. NOTE: See Land and Hold Short Operations Section.
WEATHER DATA SOURCES: ASOS (573) 499-1400:
COMMUNICATIONS: CTAF 119.3   ATIS 128.45 (1300-0300Z‡)     UNICOM 122.95
    FSS (COU) on arpt 122.65 122.2 TF 1-800-WX-BRIEF. NOTAM FILE COU.
® MIZZU APP/DEP CON 124.375
    TOWER 119.3 (1300-0300Z‡)       GND CON 121.6
AIRSPACE: CLASS D svc 1300-0300Z‡ other times CLASS E.
RADIO AIDS TO NAVIGATION: NOTAM FILE COU. VHF/DF ctc FSS. OTS indefinitely.
    HALLSVILLE (L) VORTAC 114.2    HLV      Chan 89   N39°06.81' W92°07.69'    188  18.2 NM to fld. 920/6E.
    (L) VOR/DME 110.2   COU        Chan 39   N38°48.65' W92°13.10'    at fld. 883/3E.   HIWAS.
    ZODIA NDB (LOM) 407   CO       N38°43.00' W92°16.11'    018° 6.5 NM to fld. Unmonitored when twr clsd.
    ILS/DME 110.7   I-COU          Chan 44   Rwy 02   LOM ZODIA NDB.   LOM unmonitored when twr clsd
```

Fig. 6-3. *The A/FD confirms the unicom frequency of 122.95, but note that the CTAF for traffic is 119.3—not the unicom frequency—when the tower is closed.*

tower is closed, the locally based FSS—not unicom— provides weather-related data, the favored runway, and reported traffic on the 119.3 CTAF frequency.

As with multicom, another source of airport data is the *Airport/Facility Directory*. Figure 6-3 illustrates what the *A/FD* says about Columbia Regional Airport.

But back to the uncontrolled airport with unicom. Unicom's major drawback is that it might not be in operation when you need it. Maybe it's closed; maybe the person in charge has gone to lunch, is gassing an airplane, is on the phone, or just doesn't hear your call over the blare of rock music and the hangar talk of gathered pilots, or maybe he just doesn't want to be bothered. The vast majority of unicom operators are conscientious businesspeople who want to render a service; a few, however, merely consider unicom an interruption of more rewarding pursuits.

The service can thus be limited. As a vehicle for keeping others informed and aware of your presence in the vicinity, however, unicom provides a safety measure that every pilot can and should use.

CONTACTING UNICOM

At airports with unicom but no tower or FSS, the unicom frequency is almost always 122.7, 122.8, or 123.0. These are the frequencies for airport advisories at uncontrolled airports. Let's say, then, that you're going into the Lee C. Fine Memorial Airport illustrated in Fig. 6-1. The call sign, or correct radio address, for Fine is "Kaiser," so 10 or 15 miles out, you tune to 122.8 and listen. Just as you did with multicom, you monitor the frequency to see what you can learn from aircraft that might be in the pattern or what information Kaiser unicom might be relaying to others that have already called in. If you can pick up the winds, favored runway, and so on merely by eavesdropping, all the better. You can spare the airwaves that one transmission.

Let's assume, though, that all is silent as you approach Kaiser. At least 10 miles out, the initial call to obtain the field advisory should go like this:

You: Kaiser unicom, Cherokee One Four Six One Tango.

Unicom: *Cherokee One Four Six One Tango, Kaiser unicom.*

You: Kaiser, Cherokee Six One Tango is ten miles south at four thousand, landing Kaiser. Request field advisory.

Unicom: *Cherokee Six One Tango, wind two three zero at one zero, variable. Favored runway is Two One. Reported traffic two Cessnas in the pattern.*

You: Roger, Kaiser. Thank you. Six One Tango.

A word of caution here: If you're monitoring unicom or listening to other aircraft as they report their positions in the pattern, be sure you're getting the right information from the right field. To illustrate what we mean: In one area of Kansas, at least four uncontrolled airports are within 25 miles of each other, all with the same 122.8 CTAF. Obviously, you don't have to have much altitude to hear all four (plus a few others more distant), so you could be picking up the winds and traffic at Airports B, C, or D when you were intending to land at Airport A, or vice versa. That's why it's essential to begin and end your transmissions with the name of the airport, just as you did with multicom.

At this point, several options are open to you. Your only intention is to land and tie down. Then your response is simply "Roger, Kaiser. Thank you. Six One Tango."

But maybe you want a taxi. Then it's "Roger, Kaiser. Would you call us a taxi to take us to Zandu Products?"

Or you need a mechanic: "Roger, Kaiser. Would you have a mechanic available? Our oil temperature is running high."

Or you'd like unicom to make a phone call: "Roger, Kaiser. Would you call 555-5678 and advise Mr. Schwartz that his party will be landing in about 15 minutes?"

But what if unicom doesn't respond to your first call? Try to rouse the operator a couple of more times. If you still get no response, make a blind call on 122.8 to any aircraft that might be in the pattern:

"Any aircraft at Kaiser, this is Cherokee One Four Six One Tango. Can you give me a field advisory?"

If someone answers, fine. You can then proceed to enter the pattern. Otherwise, it would be smart to fly over the field for a wind tee or sock check. The call, then, is the same as with multicom:

Kaiser *Traffic,* Cherokee One Four Six One Tango is ten south at four thousand. Will cross the field at two thousand five hundred for wind check, landing Kaiser.

Note that "Traffic" in this example is italicized. That's for emphasis, because once you have a field advisory from unicom or by other means, the unicom operator is out of the picture. Consequently, all position reports and flight intentions are now directed to other aircraft in the pattern or in the vicinity. Thus the calls are addressed to "(Blank) Traffic," not "(Blank) unicom."

A SIMULATED LANDING AND DEPARTURE WITH UNICOM

With or without an initial unicom contact, all calls follow the models we illustrated in the multicom chapter. However, to be sure that the routine transmissions are clear, let's go through them again, but without repeating the various conditions and observations.

Before Entering the Pattern

Case 1: You received no field advisory from any source, so you fly over the field to determine the wind direction:

> Kaiser Traffic, Cherokee One Four Six One Tango over the field at two thousand five hundred. Will enter left downwind for Runway Two One, full stop, Kaiser.

Case 2: You received the field advisory from unicom or another aircraft:

> Kaiser Traffic, Cherokee One Four Six One Tango entering left downwind for Runway Two One, full stop, Kaiser.

> An admonition I must repeat: Keep your eyes open. Yes, unicom might have said that two Cessnas were reported in the pattern, but do you know that they constitute the only traffic? Has someone else shown up with no radio, a radio that hasn't been turned on, or who hasn't bothered to report his presence? This is an uncontrolled airport, so your most effective life preserver might be healthy skepticism liberally sprinkled with vigilance. Think back to some of those cases in Chap. 2!

Turning Base and Final

> Kaiser Traffic, Cherokee Six One Tango turning left base for landing Two One, Kaiser.

> Kaiser Traffic, Cherokee Six One Tango turning final, landing Two One, Kaiser.

Down and Clear of the Runway

> Kaiser Traffic, Cherokee Six One Tango clear of Two One, Kaiser.

Predeparture

Leaving a unicom airport is essentially the same as with multicom, except that a call to unicom can give you the wind, favored runway, and reported traffic. This call assumes, of course, that you haven't obtained the information from a personal visit with the unicom operator—which might not always be practical because you could be at one location on the field while the FBO is somewhere else, perhaps hundreds of yards away. Assuming you haven't talked with the operator, the call is simply:

> Kaiser unicom, Cherokee One Four Six One Tango at (state your location on the field). Request airport advisory.

Once you have the advisory, and while still stationary or taxiing slowly, get on the air again and tell the local traffic what you're doing—or going to do:

Kaiser Traffic, Cherokee One Four Six One Tango taxiing to (or back-taxiing on) Runway Two One, Kaiser.

Departure

Kaiser Traffic, Cherokee One Four Six One Tango taking Two One, departing to the east, Kaiser.

After Takeoff

Kaiser Traffic, Cherokee Six One Tango clear of the area to the east, Kaiser.

Touch-and-Gos

As with multicom, communicate your intentions before taking off:

Kaiser Traffic, Cherokee One Four Six One Tango taking Two One, closed pattern for touch-and-gos, Kaiser.

On downwind:

Kaiser Traffic, Cherokee Six One Tango turning downwind for Two One, touch-and-go, Kaiser.

Turning base:

Kaiser Traffic, Cherokee Six One Tango turning base for Two One, touch-and-go, Kaiser.

Turning final:

Kaiser Traffic, Cherokee Six One Tango turning final for Two One, touch-and-go, Kaiser.

Landing or Departing the Pattern After Touch-and-Gos

When you're finished practicing and want to land, advise the traffic accordingly on downwind, base, and final:

Kaiser Traffic, Cherokee Six One Tango downwind for Two One. Full stop, Kaiser.

And so on.
If departing the pattern, the following will keep other traffic informed:

Kaiser Traffic, Cherokee Six One Tango departing the area to the east, Kaiser.

And a few minutes later:

Kaiser Traffic, Cherokee Six One Tango clear of the area to the east, Kaiser.

Now stay tuned to the Kaiser CTAF until you're 10 miles or so from the airport. This will help alert you to possible inbound traffic that might be in your general line of flight. Also, if you happen to hear an inbound pilot calling Kaiser unicom for a field advisory, don't jump in and volunteer the information, as suggested in the multicom chapter. Let unicom provide the information. However, if unicom doesn't respond after a number of efforts, it's proper to help the other pilot, based on what you knew a few minutes ago:

> Aircraft calling Kaiser unicom, Cherokee Six One Tango just departed Kaiser. Favored runway is Two One, winds about one niner at two zero, Kaiser.

A Thought for Instrument and GPS-Equipped Aircraft and Pilots

This may be old hat for most of you, but a thought struck me recently when I was riding with an experienced pilot in his GPS-equipped Cessna enroute to Abilene, Kansas, a small but active unicom-only airport. A couple of attempts to reach the unicom operator for a field advisory failed, so about 15 miles out, the pilot made his initial position report on the unicom frequency. Using the approach-plate fixes, he identified his position as "Over GPS Ezpix for landing Abilene." A few minutes later came his second report at "GPS Adela," the third at "GPS Ikehu," with the added "...landing Three Five, Abilene." And so on until we were on the final approach.

Now there was absolutely nothing out of order with any of the several position reports, except maybe this: If I were a local or transient VFR pilot with no instrument rating, no GPS knowledge or equipment, would I have any clue as to the position of this reporting aircraft? I think not. Instead, I'd be searching the skies trying to locate the traffic. I might be reasonably sure it was south of the field, but that could only increase my concern if I, too, were in that area and planning to land on Runway 35.

Visualizing this scenario, it seemed only logical to conclude that position reports based on approach plate fixes should always include the caller's altitude, geographic location of the fix relative to the airport (as north, south, southeast, or whatever), and miles out from the airport. For example: After contacting unicom for winds and favored runway, the initial call to local traffic would go something like this:

> Abilene traffic, Cessna 1234 Alpha over GPS Ezpix, fifteen southeast at three thousand five hundred, landing Three Five Abilene.

Then a few minutes later:

> Abilene traffic, Cessna 34 Alpha at GPS Adela, ten south at two thousand for straight-in approach Three Five Abilene.

And so on.

As I've said enough times, effective communication is sharing information so that there is a commonality of understanding. As the listener to position reports, I shouldn't have to wonder where my traffic is. I should know. I should have no doubts.

I do wonder, though, how often VFR pilots have been left in doubt at both controlled and uncontrolled airports when position reports from other aircraft have been basically limited to just the approach plate-reporting fixes. Such identifications would be meaningful to the listening IFR pilot (if he or she were familiar with the area or had the applicable plate at hand), but what about all the other pilots who are also nearing the field or are in the pattern or about ready to take off? Has the communication been complete, or do the listening pilots only know that there's an airplane out there over such and such fix? But where is that fix? How far is it from the airport? What is the pilot's altitude? What are his or her intentions?

A lot of people might have a lot of questions. If so, if there are uncertainties, we'd have to say that the originating caller hasn't really communicated...and that failure could conceivably be the birth of an unnecessary, perhaps serious, situation.

CONCLUSION

Flying around an uncontrolled airport presents many opportunities for unwanted confrontations. Whether multicom or unicom, the potential is the same. All I can do is urge caution and compliance with the standard radio procedures.

To quote the FAA:

> ...increased traffic at many uncontrolled airports require[s] the highest degree of vigilance on the part of pilots to see and avoid aircraft while operating to or from such airports. Pilots should stay alert at all times, anticipate the unexpected, use the published CTAF frequency, and follow recommended airport advisory practices.

Whether you are all by yourself in the pattern or one of several pilots, follow the radio procedures we've outlined. Coupled with a swivel neck and alert eyes, sharp radio technique makes the air a lot safer for everyone—especially you.

7

Flight Service Stations and Radio Communications

AS CLOSE AS THE PHONE—OR, LESS LIKELY THESE DAYS, MAYBE A short walk—is the pilot's supermarket of information and assistance. No, the Flight Service Station can't do everything, but when it comes to flight planning, weather, airport advisories, or almost everything, the FSS is a storehouse of aids that pilots need, use, or should use. If you're merely going to shoot a few touch-and-gos at the local aerodrome or wander around within a 25-mile radius of home base, the need for FSS services might be limited. But a venture farther from home demands at least a brief phone call.

For example, you're going to a field 30 miles or so away. A gusty crosswind blows you off the runway on landing and you wipe out a gear or catch a wingtip. Did you check the weather before taking off, and was your N-number recorded? If not, you might have to foot the repair bill out of your own pocket. Some aircraft insurance policies are invalid unless you can substantiate that you had taken that simple preflight precaution. The FSS was there, but you didn't use it.

THE SERVICES OFFERED: IN SUMMARY

An FSS can offer you:

- Weather information—local, enroute, and terminal, including sky conditions, winds, temperatures, dewpoints, icing, frontal activity, trends—you name it, the FSS has it.

- PIREPs (pilot reports) of conditions not easily determined from available charts or data.

- Flight plans—filing, changing, extending, and closing.

- Airport advisories for the airport where the FSS is located (when there is no tower on the airport or the tower is closed).

- Airport information (excluding traffic information) for airports with no FSS, but having a Remote Communications Outlet (RCO) and a weather observer.

- Status of restricted areas and Military Operations Areas (MOAs).

- NOTAMs—Notices to Airmen regarding airport conditions, hazards, etc.

Add one more. A sense of security, if you want to call it that. Thirty minutes after your estimated time of arrival, if you haven't extended or closed your flight plan, the FSS is on the phone checking your whereabouts. It's comforting to know that somebody down there is watching you.

With the library of help and information the FSS can offer, let's start at the beginning and follow a rough sequence of how you might make use of the various services.

FSS CONSOLIDATION

With economy and improved service the intent, the FAA has closed many of its approximately 300 Flight Service Stations and consolidated them into 61 Automated Flight Service Stations. As you might imagine, this move, while practical in many ways, was met with considerable opposition by general aviation pilots. Those in Alaska were particularly vocal, along with a significant number in the lower 48 where local weather characteristics and aviation safety warranted on-site—not remote—weather experts.

As the result of strong lobbying, the FAA restudied several of the potential closings and concluded that some 30 were located in "significant weather areas" that justified their retention. These are classified as "Auxiliary Flight Service Stations" (XFSSs). Manned by qualified weather observers, the XFSSs can effectively brief pilots on local conditions, provide airport advisories, and open or close flight plans. For more distant briefing purposes, however, since the XFSS is not automated, the on-duty specialist has to obtain enroute weather information from the area AFSS by phone or Direct User Access Terminal (DUAT)—a service that is equally available to any qualified pilot who has access to a computer.

So as of today, the number of FSSs, the vast majority of which are automated, is about one-fourth of the 300-plus that, until about 1992, were located at so many airports around the country. But, stayed tuned. The figures could change again.

What AFSSs Mean to the Pilot

The major advantage of an AFSS is, of course, its automation. Through it, the specialists have more data at their fingertips, the services have been expanded, and the pilot can obtain a wider range of information by accessing the automated menus available by telephone.

The disadvantages of consolidation? Well, one is the occasional difficulty in reaching a briefer, especially when IMC (Instrument Meteorological Conditions) exist. Also contributing to busy telephone lines, according to AFSS specialists, is caller disorganization and unpreparedness—a subject I'll address in a moment.

A second disadvantage is the absence of computer terminals in the XFSSs. This means, other than for local weather, that in-person briefings are about what the pilot could obtain by calling the appropriate AFSS himself.

In sum, though, the pluses of consolidation, so say many pilots, outweigh the minuses. If there are no incoming phone delays, briefings are faster and more thorough, flight plan filing is almost instantaneous, and many more automated services are available through pilot menu accessing.

OBTAINING A PREFLIGHT BRIEFING

(*Note:* Henceforth, for simplification, I'll refer to Flight Service Stations, including those that are automated, simply as "FSS"—not "AFSS.") Since the majority of FSS contacts are by phone, it's probably timely at this point to review the pilot's responsibilities when obtaining a weather briefing. Simply said, those responsibilities are the following:

1. Planning the flight (that's obvious)
2. Organizing the information the specialist will need to give you the briefing
3. Knowing the sequence the briefer will follow in giving you the information
4. Being prepared to copy the information as it is given

According to FSS specialists, one of the principal causes of telephone-answering delays is the pilot's lack of preparation, the disjointed sequence of information given, and the pilot's uncertainty as to what he or she really needs to know. Combined or in isolation, these deficiencies only extend the phone call and keep others in one of those annoying holding patterns. As with air communications, don't ramble and consume the briefer's valuable time, as this individual was able to do:

FSS: *Good morning. Lake City Flight Service.*

Pilot: Hi. I'm trying to get to Mountain Town. How are things between here and there? [Let's say Mountain Town is 300 nautical miles from Lake City.]

> *FSS:* *When do you plan to leave, sir?*
>
> **Pilot:** Oh, in about an hour—say ten o'clock.
>
> *FSS:* *Will you be VFR?*
>
> **Pilot:** That is Roger. [That shows he's got the lingo down.]
>
> *FSS:* *And at what altitude do you intend to fly?*
>
> **Pilot:** I guess that depends on what you tell me about the winds. How're they running?
>
> *FSS:* *Well, let's see. At three thousand they're at two nine zero at two zero; six thousand, two four zero at three zero; nine thousand, two two zero at three five—generally, southwest to west.*
>
> **Pilot:** And what's the ceiling? CAVU all the way? [That's an I-know-what-I'm-talking-about acronym for "ceiling and visibility unlimited."]
>
> *FSS:* *No. Lake City is reporting three thousand scattered, eight thousand broken, Midpoint Junction is two thousand five hundred broken, and Mountain Town is four thousand overcast.*
>
> **Pilot:** Kinda marginal VFR in spots, isn't it?
>
> *FSS:* *Yes, sir.*
>
> **Pilot:** What's the forecast for later today?
>
> *FSS:* *About the same conditions until twenty-one hundred Zulu.*
>
> **Pilot:** Twenty-one hundred Zulu—that's three o'clock tomorrow…
>
> *FSS:* *No, sir, twenty-one hundred is fifteen hundred local; and local is three P.M. today.*
>
> **Pilot:** Oh, yeah. I always get confused about whether to add or subtract. Well, what else can you tell me?
>
> *FSS:* *…Excuse me, sir. I have another call. Stand by.*

If the specialist is lucky, a telephone failure will now ensue.

Types of Briefings

Three types of briefings are available: Outlook, which provides trends 6 hours or more in advance of the planned flight departure; Standard, the normal full-length briefing; and Abbreviated, usually a capsulized update of a previously obtained Standard briefing. Because it's the most common, what follows focuses on the Standard.

Preparing for the Briefing

First, go through the normal flight planning steps—route of flight, heading, probable ground speed, proposed flight altitude, and the rest.

Second, complete the entire FAA Flight Plan form (Fig. 7-1).

Third, call the FSS. Once the briefer is on the line, the Flight Plan form now comes into play. Slowly and clearly enough so that the briefer can digest what you're saying, read off your entries in Blocks 1 through 10 on the form. Then stop. While you're reading, the briefer is typing the information into his computer terminal and, in effect, recording your flight plan. If the weather then permits a "go" decision, the plan is already in the computer. Following the briefing, assuming the flight is still "go," the briefer will ask for the rest of the information on the form—Blocks 11 through 18.

Considering what the briefer is doing, be sure to communicate the Block 1 through 10 elements in the exact sequence you have recorded them on the Flight Plan form. In other words, in this order:

1. VFR or IFR
2. Aircraft Identification
3. Aircraft Type/Special Equipment letter code (See *AIM*, Section 5)
4. True Airspeed
5. Departure Point
6. Proposed Departure Time
7. Cruising Altitude

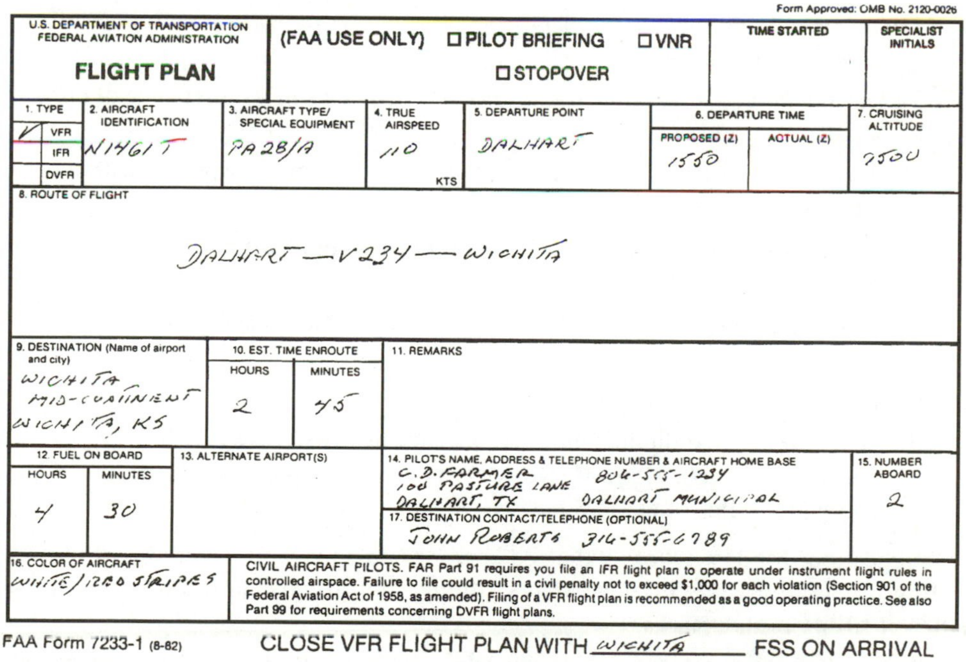

Form Approved: OMB No. 2120-0026

U.S. DEPARTMENT OF TRANSPORTATION FEDERAL AVIATION ADMINISTRATION

FLIGHT PLAN

(FAA USE ONLY) ☐ PILOT BRIEFING ☐ VNR ☐ STOPOVER | TIME STARTED | SPECIALIST INITIALS

1. TYPE: ✓ VFR / IFR / DVFR
2. AIRCRAFT IDENTIFICATION: N1461T
3. AIRCRAFT TYPE/SPECIAL EQUIPMENT: PA 28/A
4. TRUE AIRSPEED: 110 KTS
5. DEPARTURE POINT: DALHART
6. DEPARTURE TIME: PROPOSED (Z) 1550; ACTUAL (Z)
7. CRUISING ALTITUDE: 7500
8. ROUTE OF FLIGHT: DALHART—V234—WICHITA
9. DESTINATION (Name of airport and city): WICHITA MID-CONTINENT WICHITA, KS
10. EST. TIME ENROUTE: HOURS 2; MINUTES 45
11. REMARKS
12. FUEL ON BOARD: HOURS 4; MINUTES 30
13. ALTERNATE AIRPORT(S)
14. PILOT'S NAME, ADDRESS & TELEPHONE NUMBER & AIRCRAFT HOME BASE: C. D. FARMER 806-555-1234 100 PASTURE LANE DALHART, TX DALHART MUNICIPAL
15. NUMBER ABOARD: 2
16. COLOR OF AIRCRAFT: WHITE/RED STRIPES
17. DESTINATION CONTACT/TELEPHONE (OPTIONAL): JOHN ROBERTS 316-555-6789

CIVIL AIRCRAFT PILOTS. FAR Part 91 requires you file an IFR flight plan to operate under instrument flight rules in controlled airspace. Failure to file could result in a civil penalty not to exceed $1,000 for each violation (Section 901 of the Federal Aviation Act of 1958, as amended). Filing of a VFR flight plan is recommended as a good operating practice. See also Part 99 for requirements concerning DVFR flight plans.

FAA Form 7233-1 (8-82) CLOSE VFR FLIGHT PLAN WITH WICHITA FSS ON ARRIVAL

Fig. 7-1. An example of a properly completed VFR Flight Plan form. Blocks 1 through 10 should be read to the FSS briefer at the start of the weather briefing call.

8. Route of Flight
9. Destination
10. Estimated Time Enroute

Plus: Type of briefing desired: Standard, Abbreviated, or Outlook.

The Briefing Sequence

Once the briefer has the data in Blocks 1 through 10, he or she will provide the following information in this sequence:

1. *Adverse Conditions:* if any.

2. *VFR Flight Not Recommended:* If conditions appear unfavorable for VFR flight, it's at this point that the briefer will suggest "VFR is not recommended." Remember, though, the briefer is not a controller and has no authority to approve or disapprove a flight. So the go/no-go decision is yours. The fact that VFR was not recommended, however, is noted on the Flight Plan form at the very top in the small box labeled VNR (VFR Not Recommended) and it's wise to heed the briefer's suggestion. She or he is the professional. Assuming, though, that weather is no problem, the briefing continues.

3. *Synopsis:* A brief summary of weather systems that might affect the proposed flight.

4. *Current Conditions:* A summary from all sources of weather conditions applicable to the flight.

5. *Enroute Forecast:* A summary of forecast conditions in the sequence of climb out, enroute, and descent.

6. *Destination Forecast:* The forecast of significant changes expected within one hour before and after the ETA.

7. *Winds Aloft:* The forecast winds at the proposed flight altitude and, if requested, temperatures at that altitude.

8. *NOTAMs (Notices to Airmen):* Unpublished NOTAMs that could affect the flight.

9. *ATC Delays:* Any that might affect the proposed flight.

10. *Upon Request:* Additional information, such as active MOAs (Military Operations Areas), active MTRs (Military Training Routes), and the like, within 100 miles of the FSS facility.

Suggestion: Jot down those italicized headings on a piece of paper as part of your briefing preparation. You'll know what's coming and in what order. It's then just a matter of filling in the spaces.

Now, if necessary, go back and replot your flight, based on winds or other conditions the briefer may have given you that would affect your initial plan. If the changes are minor, such as a little longer or shorter enroute time, just communicate the new estimated arrival time to the FSS when you open the flight plan. If major alterations are necessary (route deviations, unanticipated intermediate stops, or the like), a call back to the FSS would probably be in order. It's easier to make significant revisions by phone than over the radio.

Just one more thing: Unless you already know it, ask the briefer to tell you what FSS frequency to use when opening your flight plan. The reason for that will be apparent in the next section.

There's a lot more to the FSS than I've indicated here—such as "Fast File," Flight Watch, Pilot Reports, enroute position reports, or FSS help when lost or disoriented. Some of these will come into the picture later. For now, though, the concentration is on the briefing service and what you, the pilot, can do to take full advantage of what the FSS has to offer. Believe me, the facility is there to help, and help it will when the pilot on the other end of the phone organizes his material, knows what he wants, asks pertinent questions, and hangs up. It's that simple.

A FEW WORDS ABOUT FSS FREQUENCIES

Four points about radio calls to an FSS follow:

- The FSS Inflight Specialist (not the briefer) with whom you would normally have radio communications can have up to 48 individual or duplicated frequencies from many remote locations to monitor. These are all displayed on a lighted panel at this position, and next to each light is a switch by which the specialist accesses a particular frequency. Then, when a pilot calls the FSS, the panel light identifying the frequency he or she is using starts to blink. The only problem, from the specialist's point of view, is that the light continues to blink only while the pilot's mike button is depressed. Once the button is released, the blinking stops and the light goes out.

    The point is this: The specialist might have heard the radio call, but if his attention was directed elsewhere and he did not see the light while it was blinking, he wouldn't know which frequency to activate. Consequently, it's important that you include in the initial call *both* the frequency you're calling on *and* your location. The location or airport name is important because two or more geographically separated airports could be assigned the same frequency.

- The following frequencies never appear on the sectional chart because they are universal and permit access to any FSS:
    121.5—for emergency use only
    122.0—"Enroute Flight Advisory Service" (EFAS): the discrete frequency for enroute weather updates and flight guidance
    122.2—for routine communications

- Beyond the unprinted "universal" frequencies, many others appear on the sectional chart, particularly those associated with Remote Communications

Outlets and VORs (Very High Frequency Omnidirectional Ranges). Just one example is the VOR illustrated in Fig. 7-2.

The thin rectangular box identifies the "Lamoni" VOR (Fig. 7-2), which can be received in flight on 116.7 or Channel 114. Next is the audible Morse code signal for "LMN"—a signal you should listen to carefully to be sure that you have tuned in the right VOR. The small square in the lower right corner of the box indicates that continuous HIWAS (Hazardous Inflight Weather Advisories) are broadcast over this particular VOR outlet. "Fort Dodge," under the rectangle, identifies that as the FSS responsible for this geographic area.

Finally, the "122.1R" above the rectangle is the frequency over which Fort Dodge can receive transmissions. Thus, to contact the FSS over the VOR, you would call on 122.1 and listen on 116.7. Addressing the call would go like this:

Fort Dodge Radio, Cherokee One Four Six One Tango at Lamoni on 122.1, listening 116.7.

When first calling an FSS, use "Radio" in the address, as in the above example, not "Flight Service." The only time "Radio" is *not* used is when you want to contact Flight Watch for enroute weather reports or updates. I'll address the matter of Flight Watch, though, in a couple of minutes.

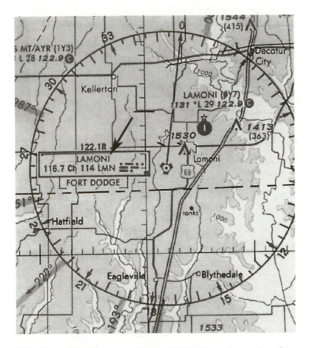

Fig. 7-2. The Lamoni, Iowa, VOR identifies other frequencies that can be used to contact the FSS in Fort Dodge, Iowa.

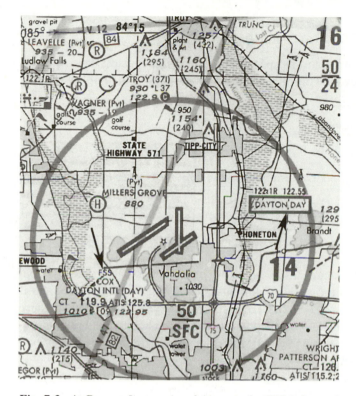

Fig. 7-3. *At Dayton International Airport, the FSS is located on the field and can be reached on 122.55, along with the other standard frequencies.*

OPENING THE FLIGHT PLAN BY RADIO

To open a flight plan by radio, four situations can exist. The radio phraseology in each, however, is basically the same, with only minor variations, depending on the airport and the location of the FSS.

Situation 1: FSS on the Departure Airport Property

You're at Dayton International Airport (Fig. 7-3). The "FSS" just above the airport name (southwest of the airport in the figure) and the heavy lined rectangle to the east of the airport symbol indicate that the Dayton FSS is located on the field and can be reached on 122.55, along with the other common FSS frequencies. So, unless you had been advised otherwise, make this call to open the flight plan on the 122.55 frequency.

That's well and good, but how should the call go? Simple.

Ground Control has cleared you to taxi, let's say, to Runway 36 for a VFR departure to Cleveland. You taxi out, complete the pre-takeoff checks, and are ready to open the flight plan. Keep in mind, though, that you're still under the jurisdiction of Ground Control, so you advise them that you're temporarily leaving that frequency:

> **You:** Dayton (or "Cox" or "International") Ground, Cherokee One Four Six One Tango leaving you to go to Flight Service.

Be sure to wait for Ground Control's approval to change frequencies; then switch to 122.55 or whatever frequency the FSS briefer had given you.

> **You:** Dayton Radio, Cherokee One Four Six One Tango on 122.55, Dayton (or "Cox" or "International").

Say nothing more now until the FSS acknowledges your call.

> **FSS:** *Cherokee One Four Six One Tango, Dayton Radio. Go ahead.*

> **You:** Radio, Cherokee Six One Tango. Request that you open my flight plan to Cleveland at fifteen ten Zulu?

Three points are to be made:

1. Make this call when you are still in the taxiway runup area—not at the taxi hold line. The reason is simple: You may have difficulty contacting Flight Service, and if you're at the hold line, other aircraft might be right behind you, ready to go, but you have them blocked while vainly trying to open your flight plan. Don't taxi to that hold line until you are really ready to leave the moment the tower clears you for takeoff. This principle applies in every case—not just here.

2. Flight Service uses Greenwich, England, time, which is Coordinated Universal Time, also referred to variously by pilots as "UTC," "Zulu time," "Zed" (French, for the letter "Z"), or even the now out-of-date GMT, for Greenwich Mean Time. Whatever the terminology, however, the flight plan and operational references to clock times should be stated in UTC times.

3. Once you're in touch with the FSS, set the intended departure time for about 10 minutes from the actual time of the call. If it's now 1500 UTC (or 10:00 A.M. local Time), ask the FSS specialist to open the flight plan at 1510 hours. The reason for these extra few minutes is that something might happen to delay your takeoff. Traffic piles up ahead of you at the hold line; there's a sudden heavy flow of inbound traffic that slows all departures, some sort of emergency blocks the active runway. Any number of things could happen, so give yourself a little leeway in case the unexpected throws your flight plan timing off schedule.

But now, picking up the call where we left off:

> **FSS:** *Roger, Six One Tango. Will open your flight plan at fifteen ten* (or the specialist might say "…ten past the hour").

> **You:** Roger, thank you. Six One Tango.

That handled, taxi from the runup area to the hold line. Come to a complete stop, change to the tower frequency, and make the call to the tower for takeoff clearance.

Situation 2: FSS Not on Airport but Has Remote Communications Outlet (RCO)

In this instance, the FSS is located elsewhere, but a Remote Communications Outlet, or RCO, is on or close to your departure airport. As an example, imagine that you're about to leave the Farmington Regional Airport in eastern Missouri (Fig. 7-4). The fact that an RCO is on or near the field is determined by the rectangular box and FARMINGTON RCO just above the airport information data. The [COLUMBIA] below the box identifies the area FSS, and the 122.3 RCO frequency is recorded on the top of the rectangle.

At Farmington, which is a non-tower-controlled airport, you first communicate your taxi and departure intentions on the 122.8 unicom CTAF, and then, after engine runup, switch to 122.3 for the FSS contact. The only difference of consequence here, versus the Dayton example, is the introduction and location identification. Despite the existence of an RCO in the immediate area, you're still talking to an FSS specialist who may be a couple of hundred miles or more away. And, remembering that the specialist could have up to 48 frequencies to monitor, with some locations having the same frequency, now is one of the times when making known your location and the frequency you're using at the outset is important. The initial wording, then, is simply this:

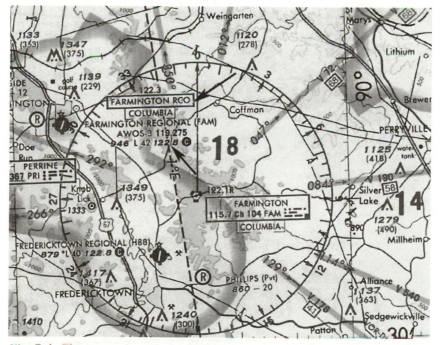

Fig. 7-4. The rectangle with the "RCO" (see arrow) indicates that an FSS Remote Communications Outlet is on or near the Farmington Regional Airport.

> **You:** Columbia Radio, Cherokee One Four Six One Tango at Farmington on 122.3.

The structure of the rest of the call is the same as the Dayton example. Of course, if local traffic permits an immediate takeoff, there's no need to add the 10 minutes to your departure time. If it's now 1500Z, just tell the specialist to open your flight at "fifteen hundred Zulu."

Once the FSS contact is completed, don't forget to go back to the 122.8 CTAF and resume transmitting the standard departure advisories to Farmington traffic.

Situation 3: No FSS on Airport; No RCO; Airport Has Adjacent VOR

You're on the ground at Waycross, Georgia (Fig. 7-5). The FSS is in Macon and there is no RCO on the field. The Waycross VOR, however, is only about 6 miles to the west and transmits on 110.2. At the same time, the "121.1R" above the VOR identifying box means that the FSS can receive on that frequency. What you do, then, is tune to 110.2 for VOR voice reception and to 122.1 on your radio. With this combination, you'll transmit on 122.1 and listen on 110.2.

> **You:** Macon Radio, Cherokee One Four Six One Tango on the ground Waycross on 122.1, listening 110.2.

[The last might not be necessary because the FSS will probably know you're listening over the VOR since you're at Waycross and transmitting on 122.1. It doesn't hurt to include the VOR frequency, though, in case there's any possible confusion.]

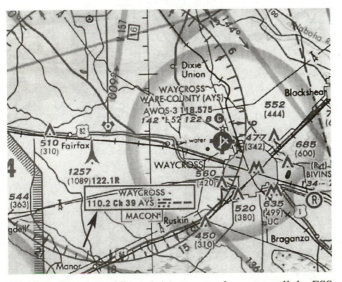

Fig. 7-5. Waycross (Georgia) is a case where you call the FSS on one frequency (122.1) and listen over the VOR (110.2).

Another possibility in this situation: It might be that the VOR is just out of radio range on the ground. Should that be the case, you'd have to take off and then make contact once airborne, with only a minor variation in the transmission:

You: Macon Radio, Cherokee One Four Six One Tango on 122.1 Waycross, listening 110.2.

FSS: Cherokee One Four Six One Tango, Macon Radio.

You: Macon, Six One Tango was off Waycross at fourteen twenty-five Zulu. Would you open my flight plan to Montgomery at this time?

And so on.

Situation 4: No FSS on the Field: No RCO; No Adjacent VOR

In cases like this, ask the FSS briefer what frequency you should use after takeoff. It might be an RCO or the nearest VOR that has voice capabilities. If it is a fairly distant RCO, though, you'll probably have to get to 3,000 or 4,000 feet agl for reception, as the typical RCO range is rather limited. Whatever the case, your radio call follows the same format as in the preceding Macon airborne example.

FILING A FLIGHT PLAN IN THE AIR

While it is possible to file a VFR or IFR flight plan while airborne, I recommend that you don't. Of course, if you're IFR qualified and run into unexpected IFR conditions, that's one thing. Otherwise, refrain. Making contact with an FSS, going through the acknowledgment, and reading off the information on the Flight Plan form just takes time—time the FSS inflight specialist could be spending on other more important issues.

In one case observed when visiting an FSS, a specialist was occupied for over 10 minutes with an IFR pilot who was air-filing and asking all sorts of questions. Meanwhile, the frequency was tied up and the specialist couldn't respond to other calls. A little thoughtfulness, plus better preflight planning, should obviate the need for most air-filings—especially VFR.

FLIGHT WATCH (ENROUTE FLIGHT ADVISORY SERVICE, OR EFAS)

You're on a VFR flight from western Tennessee to Charlotte, North Carolina. As you near Knoxville, you notice the afternoon buildup of cumulus activity ahead of you over the Appalachian Mountains. Aware of the power of these summer-spawned thunderheads, you conclude that a little information about what's happening—or is expected to happen—over your route is in order. So, Flight Watch comes into the picture. The only questions are, whom do you call, and what do you say?

First, a bit of background. Not all FSSs offer the Flight Watch service, with only 20, having been so designated. To determine those that do have EFAS in the area in which you're flying, check the inside of the back cover page of the appropriate *Airport/Facility Directory* (Fig. 7-6). For instance, in the Southeast, there are 11 FSSs, but only four—Jackson, Tennessee; Macon, Georgia; Gainesville, Florida; and Miami, Florida—provide the service.

But what if you're a hundred or so miles from one of these FSSs? How do you make radio contact? The answer lies in the remote communications outlets that extend outward from each FSS. If you'll check Fig. 7-6 again, you'll note that an outlet in the Knoxville (TYS) area connects you with the Macon FSS. Or around Florence (FLO), South Carolina, you'd be in contact with Gainesville (GNV), Florida.

Suppose, though, that you're about halfway between two of the outlets and aren't sure which FSS to call. In that case, dial in 122.0 anywhere in the country and merely address the message to "Flight Watch," your aircraft N-number, followed by the closest VOR to your position. For example: "Flight Watch, Cherokee One Four Six One Tango at Hinch Mountain VOR." That identifies your location and the FSS servicing that area will respond.

To access Flight Watch, you should normally be between 5,000 feet agl and 17,500 feet msl. Above 17,500, the FSSs around the country provide "high altitude" EFAS on frequencies other than 122.0. Also, the address should always be to the appropriate *Air Route Traffic Control Center*, even though the FSS will respond, followed by "Flight Watch," not "...Radio." This is the one time when "Radio" is not used in conjunction with an FSS.

Back to your Nashville-Charlotte flight: You know that you're in Atlanta Center's area and, by reference to the *A/FD,* that the Macon FSS provides EFAS. So, with 122.0 tuned in, you make the call:

You: Atlanta Flight Watch, Cherokee One Four Six One Tango.

FW: *Cherokee One Four Six One Tango, Atlanta (or Macon) Flight Watch.*

You: Flight Watch, Cherokee Six One Tango is over Oak Ridge at niner thousand five hundred, VFR to Charlotte. Request enroute weather advisories.

FW: *Cherokee Six One Tango, cumulus building northwest of Asheville and potential thunderstorms in the Snowbird VOR area. Asheville is clear at this time, but expect widely scattered thunderstorms by sixteen hundred local. Winds two two zero at niner thousand. Ceilings and visibility unlimited Asheville to Charlotte. Asheville altimeter three zero two five, Macon (or Atlanta) Flight Watch.*

You: Roger, Flight Watch. Thank you. Cherokee Six One Tango.

A point to remember about Flight Watch is that this is strictly a weather-related service—nothing more. Since that is its only role, never use it for routine position-reporting, filing or amending a flight plan, an emergency, or any other purpose. Those contacts should be made on 122.2 or one of the published FSS frequencies, while the emergency frequency is universally 121.5.

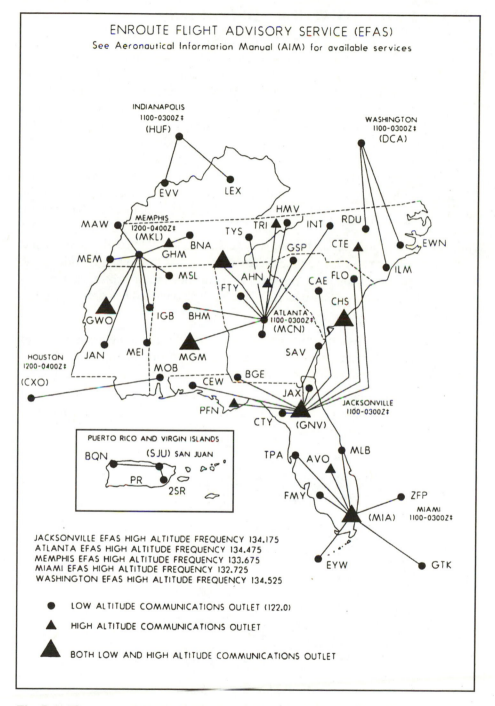

Fig. 7-6. *The nearest FSS Flight Watch to call can be determined by reference to the inside back cover of the appropriate A/FD.*

OTHER INFLIGHT WEATHER ADVISORIES

Flight Watch isn't the only source of weather information while in flight. For example, if you'll check Fig. 7-7, you'll notice the small square in the lower right corner of the Fort Dodge VOR identification box. This, as mentioned earlier, indicates the continuous recorded transmission of HIWAS (Hazardous Inflight Weather Advisory Service).

In essence, the HIWAS transcriptions contain summaries of hazardous weather within a given Air Route Traffic Control Center's area of responsibility, based on the various weather-reporting/forecasting sources. Also included are isolated thunderstorms and limited areas of low ceilings and visibilities that would not normally warrant a special weather advisory broadcast. The message concludes with: "Contact Flight Watch or Flight Service for additional details. Pilot weather reports are requested." When there are no pertinent weather advisories, a statement to that effect is broadcast hourly.

While it is beyond the scope of this book to get into meteorological discussions or to detail all of the weather-reporting vehicles, a few of the types of inflight advisories available to the pilot must be mentioned. The following are either summarized, as pertinent, in the HIWAS alert message or the specifics of each can be obtained by contacting Flight Watch or an FSS:

- AIRMETs reflect conditions that present a potential hazard to light planes and to those lacking instrumentation or equipment (as deicers or anti-icers). In effect, an AIRMET is an amendment to the area forecast.

- SIGMETs report significant meteorological developments that could be particularly hazardous to light aircraft and potentially hazardous to all aircraft. They also are included in the area forecast.

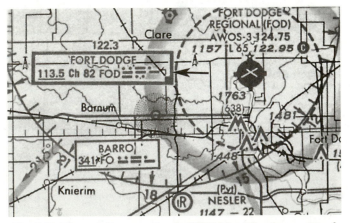

Fig. 7-7. The small square in the VOR identification rectangle (see arrow) indicates the availability of HIWAS alert messages over the VOR frequency.

All Flight Service Stations within 150 miles of the weather area broadcast SIGMETs upon issuance at 15 and 45 minutes past the hour during the first hour. Thereafter, an alert notice is broadcast at 15 and 45 minutes past the hour (for the duration of the advisory), stating that "SIGMET (name and number) is current."

If you're monitoring an FSS or a VOR and pick up only the alert notice (not having heard the initial and complete SIGMET or AIRMET), you would be wise to call either the FSS or Flight Watch for further information:

You: De Ridder Radio, Cherokee One Four Six One Tango on one two two point two.

FSS: *Cherokee One Four Six One Tango, De Ridder Radio.*

You: De Ridder, Cherokee Six One Tango. Would you read me SIGMET Delta Four?

FSS: *Cherokee Six One Tango, SIGMET Delta Four reads as follows: Flight precaution eastern Louisiana, moderate to severe turbulence in clouds seven thousand to fifteen thousand feet msl. Conditions expected to continue until zero three hundred Zulu.*

You: Roger, De Ridder. Cherokee Six One Tango.

SIGMETs report severe to extreme turbulence, severe icing, and dust or sandstorms that reduce visibility below 3 miles. In effect, they report all of the more hazardous conditions *except* thunderstorms, tornados, and hail.

Convective SIGMETs focus on these latter phenomena, and, as with AIRMETS and SIGMETS, are broadcast on the voice facilities of Flight Service Stations, as well as being available for the pilot's preflight review. Beginning daily at 0000Z, convective SIGMETs are numbered consecutively from 01 to 99, with each valid for one hour.

PIREPs (Pilot Reports) are perhaps the most significant and accurate sources of current flight conditions. Who knows better than the pilot who is there now and can report what is actually happening? Light turbulence that was forecast turns out to be moderate or severe. Unpredicted icing is experienced. Winds that were supposed to be from 240 degrees at 15 knots suddenly become brutal headwinds or helpful tailwinds at 45 knots. A flock of waterfowl finds its way into the traffic pattern. A sudden wind shear is encountered on take-off or landing. All of these unexpected events should be immediately reported.

As a general rule, submit a PIREP whenever conditions occur that were not forecast or that are actually or potentially hazardous to flight—weather-related or otherwise. Current PIREPs help forecasters, briefers, controllers, and pilots alike. (Flight Watch even appreciates reports of favorable conditions.)

What should be reported, keeping in mind the general rule stated above? Basically, unanticipated icing, clear air turbulence, wind shears, thunderstorms, ceiling or visibility changes (from those forecast), significant wind direction or velocity changes, in-flight or runway obstacles, precipitation, and anything else you encounter that was unexpected and could present a danger to other pilots.

To whom should you make the report? Depending on where you are and with whom you are in communication, the report should go to the nearest Flight Service Station, the Air Route Traffic Control Center (if you're using this facility), Flight Watch, the Control Tower, or, if at an uncontrolled airport, the local unicom. These are the sources that will make use of your report and convey it to others orally and (except for unicom operators) over the weather communications network.

An example of a PIREP communicated in flight:

You: Fort Worth Radio, Cherokee One Four Six One Tango on one two two point four five off Childress.

FSS: Cherokee One Four Six One Tango, Fort Worth Radio.

You: Fort Worth Radio, Cherokee One Four Six One Tango PIREP. Cherokee 180 is VFR 45 miles out of Childress on Victor 114 to Amarillo at 2230 Zulu. Have experienced moderate chop last 20 miles to present position. Clear of clouds at six thousand five hundred.

FSS: Roger, Cherokee Six One Tango. Keep us advised if conditions continue or worsen.

You: Roger, will do. Six One Tango.

In the PIREP, include:

• Type of aircraft
• Location
• Time (UTC)
• Conditions you are reporting
• Whether in or out of clouds
• Altitude
• Duration of conditions you are reporting

If the PIREP concerns icing or turbulence, use the approved definitions as detailed in the *AIM* (Fig. 7-8). What to you might seem to be severe turbulence could, by definition, be moderate or perhaps even light. Just be sure that you and the person on the ground are speaking the same language. Misuse of official terms because of unfamiliarity with them can result in inaccurate information to other pilots. They are either lulled into a sense of false security (and we know the dangers of that), or they are warned of conditions that don't actually exist, which could cause flight deviations, increased time enroute, and increased fuel consumption.

So you should know what you're talking about when you submit a PIREP. But even if you're not certain about how to classify the exact conditions you're encountering, according to official definition, a slightly inaccurate PIREP, when unexpected conditions arise, is better than no PIREP at all. Others will appreciate your concern for their well-being.

Turbulence Reporting Criteria Table

Intensity	Aircraft Reaction	Reaction inside Aircraft	Reporting Term–Definition
Light	Turbulence that momentarily causes slight, erratic changes in altitude and/or attitude (pitch, roll, yaw). Report as **Light Turbulence**; [1] or Turbulence that causes slight, rapid and somewhat rhythmic bumpiness without appreciable changes in altitude or attitude. Report as **Light Chop**.	Occupants may feel a slight strain against seat belts or shoulder straps. Unsecured objects may be displaced slightly. Food service may be conducted and little or no difficulty is encountered in walking.	Occasional–Less than $1/3$ of the time. Intermittent–$1/3$ to $2/3$. Continuous–More than $2/3$.
Moderate	Turbulence that is similar to Light Turbulence but of greater intensity. Changes in altitude and/or attitude occur but the aircraft remains in positive control at all times. It usually causes variations in indicated airspeed. Report as **Moderate Turbulence**; [1] or Turbulence that is similar to Light Chop but of greater intensity. It causes rapid bumps or jolts without appreciable changes in aircraft altitude or attitude. Report as **Moderate Chop**. [1]	Occupants feel definite strains against seat belts or shoulder straps. Unsecured objects are dislodged. Food service and walking are difficult.	**NOTE** 1. Pilots should report location(s), time (UTC), intensity, whether in or near clouds, altitude, type of aircraft and, when applicable, duration of turbulence. 2. Duration may be based on time between two locations or over a single location. All locations should be readily identifiable.
Severe	Turbulence that causes large, abrupt changes in altitude and/or attitude. It usually causes large variations in indicated airspeed. Aircraft may be momentarily out of control. Report as **Severe Turbulence**. [1]	Occupants are forced violently against seat belts or shoulder straps. Unsecured objects are tossed about. Food Service and walking are impossible.	**EXAMPLES:** a. Over Omaha. 1232Z, Moderate Turbulence, in cloud, Flight Level 310, B707.
Extreme	Turbulence in which the aircraft is violently tossed about and is practically impossible to control. It may cause structural damage. Report as **Extreme Turbulence**. [1]		b. From 50 miles south of Albuquerque to 30 miles north of Phoenix, 1210Z to 1250Z, occasional Moderate Chop, Flight Level 330, DC8.

[1] High level turbulence (normally above 15,000 feet ASL) not associated with cumuliform cloudiness, including thunderstorms, should be reported as CAT (clean air turbulence) preceded by the appropriate intensity, or light or moderate chop.

Fig. 7-8. How AIM describes the four turbulence-intensity levels.

EXTENDING THE FLIGHT PLAN

You're over central Kansas, enroute to Kansas City on a 3-hour flight plan. The winds, however, aren't holding up to the forecast. Either the tailwind is less or the headwinds are greater. Whatever the reason, you're not going to make the 3-hour estimate. It looks like the enroute time will be closer to 3 ½ hours, if not a little longer.

Keeping in mind that the FSS starts asking questions if you haven't closed out your flight plan within 30 minutes of your estimated arrival, you decide to extend the flight plan, based on the new ETA. If you're in the vicinity of Hays, Kansas, the call would go like this:

You: Wichita Radio, Cherokee One Four Six One Tango, on one two two point three, Hays.

FSS: *Cherokee One Four Six One Tango, Wichita Radio.*

You: Wichita, Cherokee Six One Tango over Hays at seven thousand five hundred on VFR flight plan to Kansas City Downtown, with a one five zero zero local ETA. Would like to extend the ETA to one five three zero.

> *FSS: Cherokee Six One Tango, Roger. Will extend your VFR flight plan to Kansas City to one five three zero local.*

> You: One Five Three Zero. Roger, thank you. Cherokee Six One Tango.

AMENDING THE FLIGHT PLAN IN FLIGHT

The flight is from Memphis to Kansas City, a distance of 410 statute miles. With forecast winds of about 20 knots from 270 to 290, you plan to make a pit stop at Springfield, Missouri. Along the way, though, you find the winds are much less than anticipated and that you have plenty of fuel to reach Kansas City nonstop. After passing the Dogwood VOR, 35 miles southeast of Springfield, you tune to the Springfield VOR and continue on Victor 159 toward the station.

In checking the sectional chart, you note the solid square in the lower right corner of the VOR box, indicating that the station transmits HIWAS. With this service available to you, you monitor the broadcast on the VOR frequency and find that the conditions into Kansas City justify the elimination of the Springfield stop. So you call the Columbia FSS:

> You: Columbia Radio, Cherokee One Four Six One Tango on one two two point five five, Springfield.

> *FSS: Cherokee One Four Six One Tango, Columbia Radio.*

> You: Columbia, Cherokee One Four Six One Tango is two zero miles southeast on Victor One Five Niner at eight thousand five hundred on VFR flight plan Memphis to Springfield. We have two point five hours of fuel remaining and would like to amend our flight plan and proceed direct to Kansas City Downtown, with an ETA of one four four five local.

> *FSS: Cherokee Six One Tango. Understand you have two point five hours of fuel remaining. We will amend your flight plan to show you direct to Kansas City Downtown with a one four four five ETA.*

> You: Roger. Thank you for your help. Cherokee Six One Tango.

CHECKING MILITARY OPERATIONS AREA ACTIVITY (MOAS)

Those Military Operations Areas, or MOAs, can present fairly meaningful obstacles to VFR flights, sometimes locally, depending on their proximity to an airport, and often on cross-country excursions. If it's a cross-country flight, you usually have several planning options: (1) barge through the MOA and take your chances (dumb); (2) detour around the whole thing (maybe smart but potentially expensive); (3) go only on days, at the times, and at the altitudes when, per the sectional chart's legend flap, the MOA(s) are not scheduled to be in use (okay, but this takes a lot of scheduling flexibility on your part); (4) try to go through the MOA(s) but check the activity with an FSS just before entering the area(s), then detour, if necessary (good decision).

The last being your final decision, how does the call go? It's an easy call, and this might be the expected pattern:

Let's say you're enroute from Memphis to Jacksonville, Florida. The most direct routing is via Birmingham, Alabama, over Albany, Georgia, and on to Jacksonville. Immediately east of Albany, however, lies the extensive Moody MOAs—Moody Areas 1, 2A, and 2B. To go around this airspace would add 45 minutes or so to your flight—not a major detour but a needless one in time and fuel, if it can be avoided. The answer, then, is to contact the Macon FSS about 20 miles west of the MOA to determine its current activity and the extent to which passage through it would be feasible.

You: Macon Radio, Cherokee One Four Six One Tango on one two two point two over Albany VOR.

FSS: *Cherokee One Four Six One Tango, Macon Radio.*

You: Macon, Cherokee One Four Six One Tango is ten miles northwest on Victor One Five Niner at seven thousand five hundred enroute Jacksonville on VFR flight plan. Can you advise if the Moody MOAs are hot?

FSS: *Cherokee Six One Tango, affirmative. Moody 1A and 2A are both hot at this time. Contact Jacksonville Center for advisories, frequency one three two point five five.*

You: Roger, will do. Cherokee Six One Tango.

Obviously, what Jacksonville Center tells you will dictate your actions. You're not prohibited from entering the MOA, but if there's enough high-speed activity buzzing around in there, good judgment should tell you to follow the airways and add another 45 minutes to your journey.

CLOSING OUT THE FLIGHT PLAN

One way to close out a flight plan is by telephone at your destination. The other way is via radio after you're on the ground and parked at the ramp. There can be some exceptions when an inflight close-out is in order, but the best approach is to wait until you've landed and the flight is completed.

Whichever method you choose, just be sure you do close it out! Failure to do so sets a lot of wheels in motion to track you down, and if the search turns out to be unnecessary, it doesn't sit very well with the powers that be.

If the FSS or an RCO is on the field and you select the radio, the call is simply this:

You: Jackson Radio, Cherokee One Four Six One Tango on one two two point two; Memphis.

FSS: *Cherokee One Four Six One Tango, Jackson Radio.*

You: Cherokee One Four Six One Tango is on the ground at Memphis International. Please close out my VFR flight plan from Springfield, Missouri, at this time.

FSS: *Cherokee Six One Tango, Roger. Closing out your flight plan at three five.*

You: Understand we're closed at three five. Thank you. Cherokee Six One Tango.

OBTAINING SPECIAL VFRs

First, a word about what these Special VFRs (SVFRs) are, who issues them, and conditions related to them. In essence, when the weather in the airport environment is below the minimum conditions established for VFR flight, a pilot can request that ATC issue a SVFR a clearance to enter, leave, pass through, or operate in Class D and E surface areas and some Class B and C areas. Other things being equal, ATC will grant the request, depending upon the volume of traffic and providing the SVFR flight will not delay IFR operations. Restrictions or limitations affecting SVFRs are as follows.

1. No SVFR will be issued between sunset and sunrise unless the pilot is instrument rated and his or her aircraft is equipped for instrument flight.

2. Visibility on the surface area must be at least one mile and the pilot must be able to remain clear of all clouds or of the overcast.

3. Some Class B and C airports will not approve SVFRs because of the volume of IFR traffic. Those airports are listed FAR Part 91, Appendix D, and are also identified on sectional charts by "NO SVFR" located immediately above the area containing the airport's full name and radio frequencies.

4. A complete flight plan need not be filed, but the pilot must supply enough information relative to intentions so that ATC will be able to fit the flight into the traffic flow.

5. If a tower exists on the airport, the pilot should request the SVFR through that source. Otherwise, as at a Class E airport, the request is made through the nearest operating tower, the appropriate FSS, or Air Route Traffic Control Center responsible for the area.

There are a few other SVFR regulations or restrictions outlined in *AIM,* but these are some of the more significant. So, operating within the FAR guidelines, let's illustrate the radio contacts involved in requesting SVFRs under different airport-facility conditions.

No Tower, Flight Service Station Remote

The airport is Oklahoma's Gage-Shattuck Class E (Fig. 7-9), which has a Remote Communications Outlet over which you can contact the McAlester FSS more that 200 miles to the southeast. You want to depart the Gage area and head northwest to Liberal, Kansas, but the weather is definitely below VFR limits. In deciding whether to request an SVFR, you can see that the visibility is at least a mile and the ceiling is high enough so that you could remain below the overcast or clear of any clouds. You also know from various reports that only a few miles away the standard VFR ceiling and visibility conditions [1,000-and-3 (ceiling and visibility)] exist. As things stand, though, you're groundbound—unless ATC will issue you a Special VFR clearance out of the immediate airport area. So what do you do?

What you do is prepare a flight plan and request an SVFR through the McAlester FSS. McAlester will relay the request to the appropriate ATC agency (in this case, the Kansas City Center), and then confirm the approval or denial to the pilot. Just remember that the Flight Service Station is not a controlling agency; it cannot approve or deny any operational request. Consequently, in this case, its only responsibility is to relay the SVFR request to the appropriate controlling agency and then report ATC's decision back to the pilot.

You: McAlester Radio, Cherokee One Four Six One Tango on one two two point five five, Gage.

FSS: *Cherokee One Four Six One Tango, McAlester Radio.*

You: McAlester, Cherokee Six One Tango requests Special VFR out of Gage Class E surface area for VFR northwest to Liberal via Victor Five Zero Seven. Departure time immediate upon approval of request. Six One Tango.

FSS: *Cherokee Six One Tango, stand by.*

[Pause, while the FSS contacts the Kansas City Center and then reports back.]

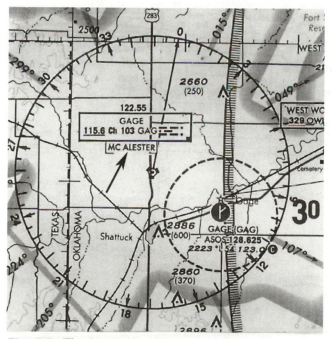

Fig. 7-9. The McAlester, Oklahoma, FSS is reached on 122.55 at the Gage, Oklahoma, airport.

> **FSS:** *Cherokee Six One Tango, ATC clears Cherokee One Four Six One Tango out of the Gage Class E surface area for northwest departure on Victor Five Zero Seven for Liberal. Maintain Special VFR conditions at or below three thousand while in the Class E surface area. Report leaving the area.*

> **You:** Roger, understand Cherokee One Four Six One Tango cleared out of Gage Class E surface area northwest, Victor Five Zero Seven, Liberal. Special VFR conditions while in E surface area. Report when clear. Six One Tango.

> **FSS:** *Readback correct, Six One Tango.*

> **You:** Roger, McAlester. Request you open my flight plan to Liberal at one seven four zero Zulu.

> **FSS:** *Will do, Six One Tango. Off Gage at one seven four zero Zulu.*

> **You:** Roger, thank you for your help. Six One Tango.

Two points here: (1) Note the use of the term "Class E surface area." Since Class E airports are already in the Class E airspace, "surface area" identifies the airport as the portion of the airspace that is of concern. (2) In a nonradar area, only one aircraft at a time is allowed in the Class E surface area when conditions are below normal VFR limits. Thus, for safety first and courtesy to others second, it is imperative to advise the controlling agency or the FSS when you are clear of the area. After establishing contact with the FSS, it's as simple as this:

> **You:** McAlester Radio, Cherokee One Four Six One Tango, Special VFR, clear of the Gage Class E surface area, northwest on Victor Five Zero Seven.

Now let's reverse the process. You want to enter the Class E surface area and land at Gage. A call to the Gage unicom, however, reveals that the weather is below VFR limits but that the 1-mile and clear-of-clouds minima exist on the field. That being the case, the FSS again comes into the picture:

> **You:** McAlester Radio, Cherokee One Four Six One Tango on one two two point five five, Gage.

> **FSS:** *Cherokee One Four Six One Tango, McAlester Radio.*

> **You:** McAlester, Cherokee Six One Tango is 20 west of Gage VFR on Victor Five Zero Seven Request Special VFR into the Class E surface area for landing Gage.

> **FSS:** *Cherokee Six One Tango, ATC has one other aircraft in the Class E surface area for landing. Stand by and remain clear of the area until further advised.*

Now do what you have to do to stay well outside the 5-mile Class E surface area until the FSS calls you again:

> **FSS:** *Cherokee One Four Six One Tango, McAlester Radio.*

You: McAlester Radio, Cherokee One Four Six One Tango.

FSS: *Cherokee Six One Tango, ATC clears Cherokee One Four Six One Tango to enter the Gage Class E surface area. Maintain Special VFR conditions at or below three thousand feet while in the Class E surface area. Report when down and clear at Gage.*

You: Roger, McAlester, understand ATC clears Cherokee One Four Six One Tango to enter Gage Class E surface area at or below three thousand on Special VFR. Report when down and clear.

FSS: *Cherokee Six One Tango, readback correct.*

You: Roger, Six One Tango.

In a few minutes, you're down and on the ramp at Gage. Now call the FSS as quickly as you possibly can so that ATC can clear another aircraft into, out of, or through the surface area:

You: McAlester Radio, Cherokee One Four Six One Tango down and clear at Gage.

FSS: *Roger, understand Cherokee Six One Tango is down and clear at Gage.*

You: That is affirmative—and thank you for your help. Cherokee Six One Tango.

Flight Service on the Airport, No Tower

In another situation, take the example of McMinnville, Oregon (Fig. 7-10)—a unicom, non-tower-controlled airport, with an FSS located on the property. The weather is below VFR minimums, so you personally visit the McMinnville Flight Service Station for a face-to-face discussion with a weather specialist. After studying conditions, it's apparent that the airport visibility is at least 1 mile and that you can remain clear of clouds within the airport airspace. It's also apparent that once clear of that airspace, you would find VFR conditions. They'd be marginal, but they'd at least be above the 1,000-and-3 minimums. Consequently, you file a flight plan and tell the specialist that once you're ready to go, you'll be requesting a Special VFR. You don't ask for that Special VFR now because it's most unlikely that the controlling ATC would issue it. The reasons (and any one of them could void the legality of an SVFR): (1) When you were actually ready to take off, the ceiling or visibility, or both, could have dropped below the SVFR limits; (2) your departure is delayed well beyond your estimated takeoff time; or (3) other aircraft, primarily IFR, might be entering or leaving the airport airspace, and an earlier-approved SVFR aircraft flying in the area could severely affect the planned flow of IFR traffic in the airspace.

Once you're ready to go, though, the contacts are the same as in the Gage example. Whether the FSS is on the field or remote really makes no difference. The actual radio procedures for entering, departing, or operating within the airspace (as doing touch-and-goes) don't vary.

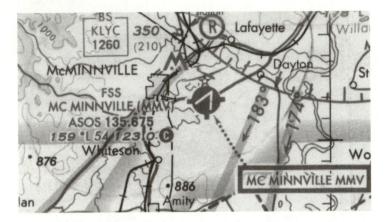

Fig. 7-10. In this instance, the FSS is located on the McMinnville, Oregon, airport property.

OBTAINING AIRPORT ADVISORIES: FLIGHT SERVICE ON THE AIRPORT, NO TOWER

When an operating Flight Service Station is located on the airport and there's no operating tower, as at McMinnville, the FSS provides the same airport advisory that unicom offers at fields with no tower or FSS, but it's more detailed and more accurate.

If you're landing at an airport with an FSS and want an airport advisory, what frequency do you use? The answer is the same at all FSS airports: 123.6. This is the standard frequency for the AAS, or Airport Advisory Service. It is not, however, the frequency for filing or closing out flight plans or for obtaining enroute weather information. Going into a field which has both Flight Service and unicom, use the latter for phone messages, taxis, and the like, and the former for the advisory service. Accordingly, and on frequency 123.6, the communication goes like this:

You: McMinnville Radio, Cherokee One Four Six One Tango on one two three point six.

FSS: *Cherokee One Four Six One Tango, McMinnville Radio.*

You: McMinnville, Cherokee Six One Tango is ten miles west of McMinnville VOR at three thousand. Request airport advisory.

FSS: *Cherokee Six One Tango, favored runway is two two, wind two four zero at one two, altimeter two niner eight seven. No reported traffic.*

You: Roger. Thank you. Cherokee Six One Tango.

KEEPING LOCAL TRAFFIC INFORMED

Keep in mind that Flight Service is not a traffic control agency at McMinnville or anywhere else. So once you enter the pattern, stay on 123.6 but address your communiques to "McMinnville Traffic," not "McMinnville Radio."

From this point on, the calls are the same as at a multicom or unicom field—entering the pattern on downwind, turning base, turning final, and on the ground and clear of the runway.

As with unicom, don't expect any response from the FSS when you make these transmissions. I've noticed on several occasions, however, particularly at the smaller or less busy airports, that the specialist does acknowledge even routine position reports, this despite the pilot's use of "Traffic" rather than "Radio." So, if the FSS does respond to or acknowledge blind transmissions, don't be surprised. On the other hand, don't be surprised if silence reigns. That's the way it should be.

Remember, too, as with unicom, that there's no requirement for you to contact Flight Service for an advisory. Not doing so is a little foolish, though. Fulfill your responsibility, but keep your eyes open and your head turning to spot any possible "Silent Sams."

TAXIING OUT AND BACK-TAXIING: FLIGHT SERVICE ON THE AIRPORT, NO TOWER

As we said in the multicom and unicom chapters, if you're going to back-taxi, complete the entire pre-takeoff check before venturing onto the runway—unless an area exists at the end of the runway for that purpose. It's simply discourteous to others and disruptive to traffic to park on the approach end while you go through the checklist. More than one landing pilot has been forced to go around because some self-centered idiot was sitting there on the runway, oblivious to the existence of all others. If there is no runup area, make the check on the ramp or taxiway, back-taxi with some speed and power, do a 180, and get going.

But to begin at the beginning: you're on the ramp, engine started, and you want an airport advisory. Tune to 123.6 and call Flight Service:

> **You:** McMinnville Radio, Cherokee One Four Six One Tango on the ramp, ready to taxi for takeoff. VFR northbound. Request airport advisory.

> **FSS:** *Cherokee Six One Tango, favored runway is One Seven, wind one niner zero at four. Altimeter three zero zero zero. Mooney reported ten west for landing.*

> **You:** Roger, taxiing to One Seven. Cherokee Six One Tango.

You're ready to go, *after* the pre-takeoff check:

> **You:** McMinnville Traffic, Cherokee Six One Tango taking One Seven, northwest departure, McMinnville.

CTAFs: Part/full-time FSS, part/full-time tower

Budget-cutting in recent years has brought about a profusion of part-time towers and part-time FSSs. As a general rule, when a tower exists on a field, even if it is closed, use the tower frequency for traffic advisories. If there is no tower but an FSS exists on the field, use 123.6 for self-announce advisories even if the FSS is closed. Figure 7-11 summarizes the various possibilities, but remember to check the *A/FD* for exceptions.

Tower Status	FSS Status	Frequency
Open tower on field	Open, closed, or none	Tower frequency
No tower on field	Open on field	AAS on 123.6
No tower on field	Closed FSS on field	Self-announce on 123.6; advisories on unicom, if available
No tower on field	No FSS on field	Unicom (if available) or multicom (check A/FD)
Closed tower on field	Open FSS	AAS on tower frequency
Closed tower on field	Closed (or no) FSS on field	Self-announce on tower frequency

Fig. 7-11. The frequencies for airport traffic advisories, depending on the existence and current status of an on-airport control tower or Flight Service Station.

IF YOU'RE LOST…

In previous editions, I discussed the ability FSSs had to provide Direction Finding (DF) fixes or DF steers to help pilots who were lost or uncertain of their position. This ability required certain receiver equipment, which was decommissioned, as of October 1, 1997, thus making the DF services, as once described, unavailable through any of the FSSs.

But, even if you're lost, all is not. Flight Service can still be of help. To make what follows logical, though, at least two conditions should prevail: One, you're on a cross-country flight over territory that is rather barren of distinguishing landmarks; two, you have not been contacting a Center for traffic advisories, so the Center knows nothing about you.

Suddenly, or perhaps gradually, you find yourself searching for landmarks that would identify your position. After a few minutes of trying to match what you see on the ground with designs and symbols on the sectional chart, you reach the uneasy conclusion that you don't know where you are. If, coupled with that conclusion, the fuel gauge needles are dropping, a bit of anxiety might be developing. (You snicker? People are still getting lost, even today.)

To be honest, though, if you have even one operating navcom on board, you should be able to orient yourself with little difficulty. Merely maintain whatever heading you're on and tune to a VOR station. Turn the navcom's OBS (Omni Bearing Selector) until the needle is centered and then draw a line on the sectional representing the radial from the VOR. Next, and while still on the same heading, tune to a second VOR and draw the radial from that station. Where the lines from the two VORs intersect, that's where you are.

But, perhaps you're a student pilot or one with very little experience, and the prospect of being lost, with diminishing fuel, creates a sweaty-palms condition. You want help from the ground. Here is where the nearest FSS can come into play. If you're not sure to which FSS you should address your call, dial in the 122.2 FSS frequency, determine the VOR station closest to your present position, and then address the call as follows:

> **You:** Any Flight Service Station, this is Cherokee One Four Six One Tango (over or near) the (blank) VOR. Request orientation assistance.

(Or words to that effect. This is not yet an emergency, so the procedures involved in *distress* or *urgency* situations (see Chap. 10 for a discussion of these conditions) are not yet applicable here.]

Once in contact with the FSS specialist, the steps he or she would follow are essentially the same as those described above: This specialist would ask you to identify a VOR, then tune to it, rotate the OBS until the needle is centered, and then report the radial from the station. The specialist would then plot that radial on a sectional chart physically located nearby in the FSS office. Following that, the specialist would ask you to do the same with a second VOR, after which he or she would plot that radial. Again, where the lines cross is where you are. After identifying your location, and based on your request, the specialist could (1) terminate the contact, (2) direct you to the nearest airport, or (3) give you headings that would send you on your way.

A question at this point: If the pilot can tune to the VORs and draw the lines on the sectional chart, why call in an FSS? Answer: There should be no need to, *if* you have the aircraft under control, *if* your avionics are working properly, *if* the visibility is reasonable, and *if* you're not having an anxiety attack. But new or inexperienced pilots might not find everything smooth and orderly with ample fuel and reliable VORs. They might want help now before a worrisome situation becomes a bona fide emergency. Far better, then, to put pride aside and get help when help can really be helpful. The alternatives might not be as pleasant to contemplate.

One other scenario (keep in mind that in these illustrations, you have not been in contact with a Center, for whatever reason; if you had, there might be no need to call on an FSS): You have one navcom, but it's either dead or is highly unreliable. Very unsure of where you are, you call the nearest FSS on 122.2, identify yourself, and explain your situation. In response, the FSS specialist will most likely tell you to stand by. He or she will then request a transponder code for you from the appropriate ATC agency (almost certainly a Center) and advise you of the same. You insert that code and confirm that fact to the FSS, after which the specialist will tell you how and on what frequency to contact the Center controller. After reaching the Center, the controller will advise you of your position and give you whatever instructions are necessary to help you become reoriented or that will lead you to the nearest airport.

THE FSS AND POSITION-REPORTING

Let's say that you're on a nonstop cross-country flight from Alpha City to Charlie City and the route takes you over or near Bravo Town. It's a 400-mile jaunt, which, based on your briefing, should take about 4 hours. So you file a flight plan with a 1200-hour departure and 1600-hour ETA. You open the flight plan by radio with FSS Alpha City at noon, and Alpha City then sends the pertinent operational data, including your 1600-hour ETA, to the FSS responsible for the area in which Charlie City is located. Once you're underway, is there any benefit to reporting your position to an FSS as you move down the road? Potentially, yes.

Suppose you're between Bravo Town and Charlie Cities. You haven't contacted any facility since leaving Alpha City, but suddenly you have a major mechanical emergency. There's no time to think about radio calls or squawking the emergency 7700 transponder code. All that matters is getting that airplane on the ground now, hopefully in one piece. So down you go, ending up in some desolate plot of landscape.

An hour or so later, the 1600 ETA comes and goes with no word from you. At 1630, your computerized flight plan alerts the Charlie City FSS that the flight plan has not been closed. A Charlie City specialist then gets on the phone to see if you actually departed Alpha City or if you had indeed landed at Charlie City but had failed to close out the flight plan.

These efforts proving fruitless, the calls and messages start flowing to determine if any facility along your route has heard from you. If not, the Search-and-Rescue (SAR) unit serving that area is alerted to a possible downed aircraft. Maybe the Air Force gets into the act. Calls are made to your family or business associates, hoping for helpful information. With nothing forthcoming, the physical search begins.

Since no one knows where you are between Alpha and Charlie Cities, the search has to cover the whole 400-mile stretch of territory. Meanwhile, you might be lying injured in a wrapped-up piece of fuselage that is no more than a speck of metal to those in the airborne SAR mission.

Time being critical in such an instance, just one enroute position report could be a lifesaver. Had you contacted the FSS serving the Bravo Town area, which, let's say, is halfway between Alpha and Charlie Cities, the search effort would have gone on as indicated, but it would have been only between Bravo Town and Charlie City rather than over the whole 400-mile stretch. That one radio call, then, theoretically doubled your chances of an early rescue. And the call is no more complicated than this:

You: Bravo Radio, Cherokee One Four Six One Tango on one two two point two at Bravo. Position report.

FSS: Cherokee One Four Six One Tango, Bravo Radio. Go ahead.

You: Bravo, Cherokee Six One Tango is ten north of Bravo Town at one four two five local, VFR to Charlie City. Six One Tango.

FSS: Roger, Cherokee Six One Tango. Over Bravo at one four two five local. Bravo altimeter two niner four five.

You: Two niner four five. Roger, Bravo. Thank you. Six One Tango.

The Bravo Town FSS now has your position and time on tape, in the event either is needed in the future.

Disregarding for the moment the safety value of position reports, the importance of closing out a flight plan perhaps becomes obvious. The same process summarized above is initiated 30 minutes after your ETA has expired if no FSS has heard from you. Obviously, it can be a costly proposition when you fail to make that last FSS phone or radio call, no matter where you may have eventually landed. Always close out your flight plan with the nearest FSS!

I'll talk more about emergency actions in later chapters, as well as the advantages of being in contact with the Air Route Traffic Control Centers and how those facilities can help you when things go wrong. Meanwhile, your friendly FSS represents, in all respects, a pretty solid insurance policy. I hope you'll agree that it's a policy worth buying when you venture forth across the airways.

8
Automatic Terminal Information Service (ATIS)

BEFORE CONTINUING, I'D LIKE TO INDICATE THAT THIS BOOK HAS A certain sequential flow or logic. Essentially, it attempts to take you from the simplest of all of radio communications at multicom and unicom airports up through the use of Approach/Departure Control and Center on a cross-country flight.

It's not always feasible, however, to maintain that exact sequence, as Chap. 7 on Flight Service Stations, illustrates. In describing the various FSS services, it seemed practical to include examples of typical inflight and enroute communication exchanges that can take place between a pilot and an FSS specialist. In that sense, the intended sequence was violated. On the other hand, for cross-country flights beyond the local airport vicinity, the FSS is typically the first contact a pilot has with the National Airspace System, because it's here where the weather is determined and the flight plan is filed.

From now on, however, and to maintain sequence, the other ATC services will be reviewed in the order that you would or could encounter them on either a local flight or a cross-country excursion. Beginning with this chapter, these facilities and services include ATIS, Ground Control, the tower, Approach and Departure Control, concluding with the Air Route Traffic Control Center. Finally, as sort of a review, I'll wrap everything up with a simulated cross-country flight that illustrates all of the radio contacts from engine start-up to final shut-down.

WHAT IS ATIS?

The Automatic Terminal Information Service, with its neat little acronym of ATIS, is hardly a stranger to most pilots—especially those acquainted with the larger controlled airports. As a source of important information, it is a vital communications tool for both the departing and arriving airman.

For those not familiar with it, ATIS is a continuous tape-recorded summary of local weather conditions, winds, and runway(s) and instrument approach(es) currently in use. Not more than an hour old, the purpose of the transmissions is to provide pilots with the essential current airport data on a radio frequency that doesn't interfere with the tower's primary responsibility to control traffic in and around the airport environment. Aided by ATIS, the tower personnel can thus concentrate on that responsibility rather than having their attention constantly distracted by the need to advise departing or arriving aircraft of the same basic airport information.

Recorded in the towers at most high activity airports, the ATIS information is transmitted over a discrete VHF frequency or the voice portion of a local navigation aid (NAVAID) such as a VOR outlet. With the discrete frequency, ATIS can usually be received up to 25,000 feet and 60 miles from the transmitter. The information, updated every hour, is identified by the phonetic alphabet of Alpha, Bravo, Charlie, etc. Should conditions change in any material way before the next scheduled update, a revision is issued and given the next phonetic alphabet designation.

DETERMINING THE APPROPRIATE FREQUENCY

If ATIS is available at a given airport, its frequency is indicated on the sectional and in the *Airport/Facility Directory.*

To illustrate from the sectional, take the Central Nebraska Regional Airport in Grand Island as an example (Fig. 8-1). In this case, the ATIS is clearly identified as 127.4. If no frequency is stated, the service is not provided, and winds, altimeter, and runway are obtained from the tower.

INFORMATION PROVIDED BY ATIS

Assuming no unusual or potentially hazardous conditions, the typical ATIS includes the following data:

- Location
- Information code (phonetic alphabet)
- Time (UTC, "Coordinated Universal Time," often stated as "Zulu")

- Sky conditions (often omitted if ceiling is higher than 5,000 feet, or is stated as "better than five thousand")

- Visibility (often omitted if visibility is greater than 5 miles, or is stated as "better than five miles")

- Temperature and dewpoint

- Wind direction (magnetic) and velocity

- Altimeter setting

- Instrument approach in use

- Current runway in use

- NOTAMs (if any)

- Information code repeated (phonetic alphabet)

A typical example:

Jackson Information Delta. One six four five Zulu weather. Four thousand five hundred broken, visibility five. Temperature four five, dewpoint three eight. Wind two zero zero at one six, gusting to two two. Altimeter three zero zero niner. ILS Runway One Five in use, land and depart Runway One Five. Advise you have Delta.

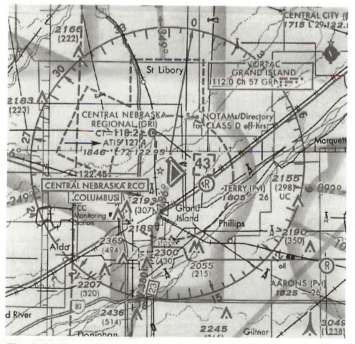

Fig. 8-1. *The ATIS frequency (see arrow) is easy to spot at tower airports.*

Depending on the facility, some ATIS tapes include departure and approach frequencies and the clearance delivery frequency. In the event of adverse weather, ATIS broadcasts include AIRMETs and SIGMETS, such as "SIGMET Charlie Two is now in effect." To determine what Charlie Two is all about, you have to contact Flight Service. ATIS does not provide all of the details.

To illustrate some of the additional information that could be included in an ATIS broadcast, let's take this example:

> Jackson Information Foxtrot. One three three zero Zulu special observation. Five hundred scattered, estimated ceiling one thousand overcast, visibility five, light rain. Temperature seven zero, dewpoint six two. Wind three two zero at seven. Altimeter two niner niner zero. VOR Delta approach in use. Left traffic. Land and depart Runway Three Five. SIGMETs Charlie Two Niner, Charlie Three Zero, and Charlie Three One now in effect. Numerous trenches along Runway Three Five. Advise you have Foxtrot.

WHEN TO TUNE TO ATIS

Departure: The engine is started; radios and beacons are on. Before any further action, tune to the ATIS frequency and listen. As it is a continuous transmission, listen twice—especially if the details of the first transmission weren't clear. Set the altimeter, and remember the information code.

Arrival: At 20 or 25 miles out, tune to the ATIS frequency and listen again—perhaps a couple of times until you have the details firmly in mind. Presumably you're not too busy at that distance from the field, so you can give attention to what's being transmitted and plan your approach to the airport accordingly.

COMMUNICATING THE FACT THAT YOU HAVE THE ATIS

Tell those who need to know that you have the current ATIS that you do indeed have it. This includes Ground Control for departure, and Approach Control or the tower when landing.

To Ground Control (when you're ready to taxi):

You: Jackson Ground, Cherokee One Four Six One Tango at Avitat with Information Foxtrot. Ready to taxi. VFR New Orleans.

GC: *Cherokee One Four Six One Tango, taxi to Runway Three Five.*

You: Roger. Taxi to Three Five. Cherokee Six One Tango.

To the Tower: You're inbound and over an identifiable checkpoint 20 miles or so from the field and this is your first arriving contact with a traffic controlling facility. The initial call, then, is simply this:

Albany Tower, Cherokee One Four Six One Tango over Smithville at three thousand five hundred for landing Albany with India.

Note that your present altitude is included in the call, along with the phonetic designation of the ATIS you have just monitored.

If the airport has an Approach Control, the structure of the message is the same:

1. N-number
2. Location
3. Altitude
4. Intentions (landing primary airport, secondary airport, or transiting)
5. ATIS information you have just monitored.

If you fail to indicate that you have the ATIS information, you'll probably be questioned: "Cherokee Six One Tango, do you have ATIS?" Or the tower or Approach might offer the basic data necessary for your entry into the pattern and for landing. That's a nice gesture, but it consumes airtime.

USING THE PHONETIC DESIGNATION

If you've received the ATIS, make it clear to the controller that you have Information Alpha, Bravo, Charlie, or whatever. That tells the controller that you are in possession of the most current and complete information.

Perhaps it's sort of classy to come out with "have the numbers," but it's misleading. It means to the controller that you have received the wind, altimeter setting, and runway only—not the full ATIS, which might include SIGMETS, NOTAMS, ceilings, visibility, and so on. Hearing "have the numbers" could cause the controller to question you further as to the extent of your information or require him to relay significant information that is included in the latest ATIS.

Another thing: Merely reporting that you "have the ATIS" is not enough. Maybe a special update was just issued. Maybe you had Information Papa 20 minutes ago, but Information Quebec has come out in the meantime. And maybe Quebec designates Runway 18 as the active runway, whereas Papa had it as Runway 26. Any number of changes might have taken place between Papa and Quebec.

To eliminate confusion and reduce airtime talk, be brief but be complete. Use the phonetic alphabet to identify the ATIS you have received. It takes no longer to be specific—and it's proper procedure. As the AIM says, "'Have the numbers'...does not indicate receipt of the ATIS broadcast and should never be used for this purpose."

WHEN UNABLE TO REACH THE ATIS

If, for any reason, you're unable to receive the ATIS or the transmission is garbled, tell Ground Control before taxiing out, or tell this to tower or Approach Control if you're just arriving. Merely conclude the call with "negative ATIS" or "ATIS garbled." The facility you're contacting will then give you the pertinent data so that you can proceed accordingly.

CONCLUSION

ATIS broadcasts exist to: (1) to provide you with the departure or arrival information you need (at worst, the information is only 59 minutes old, and is thus a reasonably accurate depiction of existing conditions), and (2) to reduce unnecessary communications, thereby permitting pilots and controllers to concentrate on their primary responsibilities—the safe operation of the aircraft and the safe direction of air traffic.

Use ATIS wherever it is available. You don't have to talk back to it. All it requires is the willingness to listen and the ability to remember. If you can't do either of those, you shouldn't be in the air in the first place.

9
Ground Control: The Airport Surface Traffic Director

MOST PILOTS FAMILIAR WITH THE BUSIER CLASS B, C, AND D AIRPORTS are well acquainted with Ground Control and the function it plays in assuring the safe operation of earthbound traffic. However, the inexperienced airman, venturing into one of these airports for the first time, needs to understand the responsibilities of Ground Control and the communications with the controller. And, may I add, so do some of the more experienced aviators. Let's just say that I've heard a lot of hesitant radio calls, incomplete calls, unprepared calls, and excessive verbal garbage—all of which reflects poorly on those who should know better.

What follows, then, will hopefully be of some value to all general aviation pilots, regardless of their levels of experience.

WHAT DOES GROUND CONTROL DO?

Ground Control is the invisible "policeman" ensconced in the tower who is responsible for the operation of aircraft and all other vehicles that are using taxiways and runways other than the active runway. Controlling the movement of those vehicles through radio communications, the tower is the authority. When the controller says "go," you go; when he says "stop," you stop; when he says "give way," you give way, and so on. If, for any reason, safety or otherwise, you can't comply with the directions, you have the responsibility to advise the controller immediately. Adherence to instructions is expected.

At almost every airport, however, some areas are not under the control of Ground: for example, the expanses of concrete used for tiedown purposes, fueling, moving aircraft to and from hangars, compass swinging, and the like. As long as these areas don't infringe on runways or taxiways, you can taxi your aircraft around there all day. Nobody will say a word. Otherwise, your intended movements must be cleared by Ground Control.

FINDING THE GROUND CONTROL FREQUENCY

The Ground Control frequency will not be found on any sectional. Where the service exists (at most tower-controlled fields), you can locate the frequency in the *Airport/Facility Directory,* as Fig. 9-1 illustrates.

One fact you can usually count on: the frequency will be 121.6, 121.7, 121.8, or 121.9. There are a few exceptions to the general rule, however. For instance, Miami International Airport uses one frequency for one set of runways and another frequency for the other runways. Memphis has one frequency for general aviation, while a second is reserved for air carriers. Consequently, to be sure of your Ground frequency, consult the latest edition of the *A/FD.*

WHEN TO CONTACT GROUND CONTROL

The following are typical situations in which you would call Ground Control. Some contacts are required; others are optional, as noted.

Taxiing for Takeoff (Required)

You have the ATIS information, you're tuned to the correct Ground Control frequency, and are ready to taxi out for takeoff now—not one, three, or five minutes from now. As I've said before, you can roam around the ramp or any uncontrolled area all you want, but don't venture out onto any taxiway until you have contacted Ground, and have received permission to taxi.

> **You:** Lincoln Ground, Cherokee One Four Six One Tango at the terminal with Information Lima. VFR Omaha.
>
> *GC:* *Cherokee One Four Six One Tango, taxi to Runway Three Five.*
>
> **You:** Roger. Taxi to Three Five. Cherokee One Four Six One Tango.

In this contact, be sure to include your location (e.g., "...at the terminal..."). If you don't, Ground will always come back with something akin to, "Cherokee Six One Tango, where are you parked?" Avoid this needless request by communicating your position in

```
ATLANTA
  DEKALB-PEACHTREE   (PDK)   8 NE   UTC-5(-4DT)   N33°52.54' W84°18.12'              ATLANTA
  1002    B    S4   FUEL 100, JET A   OX 1, 2, 3, 4   TPA (See Remarks)   LRA        COPTER
  RWY 02R-20L: H6001X100 (CONC-GRVD)    S-46, D-66    HIRL    0.4% up SW         H-4H, 6F, L-20E, A
    RWY 02R: REIL. VASI(V4L)—GA 3.9°TCH 35'. Building. Rgt tfc.                       IAP
    RWY 20L: MALSF. PAPI(P2L). Thld dsplcd 1000'. Pole.
  RWY 16-34: H3966X150 (ASPH)    S-20    MIRL
    RWY 16: RAIL. VASI(V4L)—GA 3.4°TCH 30'. Pole.
    RWY 34: REIL. VASI(V4L)—GA 3.3°TCH 39'. Trees.
  RWY 02L-20R: H3744X150 (ASPH)    S-20    MIRL
    RWY 02L: PAPI(P2L). Trees.    RWY 20R: PAPI(P2L). Trees. Rgt tfc.
  RWY 09-27: H3378X150 (ASPH)    S-20    HIRL    0.8% up W
    RWY 09: REIL. VASI(V4R)—GA 3.4°TCH 28'. Trees.    RWY 27: REIL. VASI(V4L)—GA 3.8°TCH 49'. Trees.
  AIRPORT REMARKS: Attended continuously. Heavy helicopter ops NW corner of arpt. Helipad located north of Rwy 16
    thld. Flocks of birds on or near arpt during dalgt hours. When twr clsd HIRL Rwy 02R-20L preset med ints:
    TPA—2002 (1000) single engine, 2502 (1500) multi engine. PPR for all transient military acft. Voluntary ngt
    curfew in effect from 0400-1100Z‡. Noise sensitive area all quadrants; pilots use noise abatement procedures
    prescribed by arpt director call 770-936-5440. ACTIVATE other ints and MIRL Rwy 16-34 and
    Rwy 02L-20R; HIRL Rwy 09-27; MALSF Rwy 20L and twy lgts—120.0. Flight Notification Service (ADCUS)
    available. NOTE: See Land and Hold Short Operations Section.
  WEATHER DATA SOURCES: LAWRS.
  COMMUNICATIONS: CTAF 120.9    ATIS 128.4    UNICOM 122.95
    MACON FSS (MCN) TF 1-800-WX-BRIEF. NOTAM FILE PDK.
    PEACHTREE RCO 122.1R 116.6T (MACON FSS)
    ATLANTA APP/DEP CON 119.3
    PEACHTREE TOWER 120.9 120.0 (1130-0400Z‡ Mon-Fri 1200-0400Z‡ Sat-Sun)   GND CON 121.6
      CLNC DEL 125.2
  AIRSPACE: CLASS D svc Mon-Fri 1130-0400Z‡. Sat-Sun 1200-0400Z‡ other times CLASS G.
  RADIO AIDS TO NAVIGATION: NOTAM FILE PDK.
    PEACHTREE (L) VOR/DME 116.6    PDK    Chan 113    N33°52.54' W84°17.93'   at fld. 970/02W.
    ILS 111.1   I-PDK   Rwy 20L.   Coupled apchs not authorized. Unmonitored when twr clsd.
  • • • • • • • • • • • • • • • • • • • • • • • • • • • • • • • • • • •
  HELIPAD H1: H56X56 (CONC)
  HELIPORT REMARKS: H1 perimeter lgts opr dusk-dawn.
```

Fig. 9-1. *The A/FD clearly identifies the Ground Control frequency (see arrow) at those airports where the service exists.*

the first call. By including your destination or direction of travel (which really isn't essential), you might be directed (winds permitting) to a runway more closely aligned with your departure route.

Note that in the dialogue the controller has told you to taxi *to* Runway 35. That means exactly what it says. You can go to the engine runup area, stop there for the cockpit check and runup, and then taxi to the runway hold line—but no farther. Clearance to taxi onto the active runway for takeoff comes from the tower, not Ground. No further clearance is necessary, however, to taxi from the runup area to that hold line. The fact that Ground has approved your movement to Three Five is all that is required. Another thing: While you're moving down the taxiway, once cleared, keep listening to Ground. The controller might have further instructions for you, such as:

GC: *Cherokee One Four Six One Tango, pull to the right and give way to the Bonanza taxiing in on Charlie (the taxiway).*

You: Roger, will do, Six One Tango.

Or:

> *GC:* *Cherokee One Four Six One Tango, turn left at the next intersection to let the 727 pass.*
>
> **You:** Roger, Ground, will do. Six One Tango.

Something else to keep in mind at large or busy airports is that the Ground Controller might lose track of you. In a peak traffic period, for example, he tells you to "…follow Charlie to Runway Two Seven and hold." This you do—and you hold and you hold and you hold, awaiting further word from Ground. But the controller has either lost you or has forgotten about you amidst the demands from other ground traffic. The practical thing, then, is to tell Ground when you've reached "Two Seven":

> **You:** Ground, Cherokee Six One Tango holding at Two Seven.

This does two things: First, it reattracts the controller's attention to you, and second, it advises him that you've followed his instructions. Controllers like that, especially the latter. Then when he clears you across Two Seven, be sure to acknowledge the approval:

> *GC:* *Cherokee One Four Six One Tango, clear to cross Runway Two Seven.*
>
> **You:** Roger, clear to cross Two Seven.

Soliciting Progressive Taxi Instructions (Optional)

You have the ATIS, but you're new to the airport and need taxi instructions. Ground Controllers understand that taxiway systems can be confusing to pilots, so they will give progressive (step-by-step) taxi instructions, if you ask for them:

> **You:** Lincoln Ground, Cherokee One Four Six One Tango at the terminal with Information Lima. VFR Omaha. Request progressive taxi.
>
> *GC:* *Cherokee Six One Tango, Roger. Taxi to the edge of the ramp and hold.*
>
> **You:** Roger, edge of the ramp and hold. Cherokee Six One Tango.

After proceeding as told, come to a stop, acknowledge your position, and wait for Ground to come back with further instructions.

> **You:** Ground, Cherokee Six One Tango holding at ramp edge.
>
> *GC:* *Cherokee Six One Tango, you are at Taxiway Charlie. Turn right on Charlie to Runway Two Seven and hold.*
>
> **You:** Right on Charlie. To Two Seven and hold. Six One Tango.
>
> **You:** Ground, Cherokee Six One Tango holding Two Seven.
>
> *GC:* *Cherokee Six One Tango, cleared to cross Two Seven. Turn left at the next intersection to Delta. Follow Delta to Runway One Five and hold. Contact the tower when ready.*

You: Six One Tango cleared to cross Two Seven, left on Delta to One Five and hold. Cherokee Six One Tango.

Just don't be afraid to call Ground again if you've forgotten an instruction or become confused. The folks in that glass enclosure are there to help you and to keep things running smoothly on those strips of concrete.

Taxiing in after Landing (Required)

You've touched down, and the tower has instructed you to "contact Ground point niner." (The tower might give you the complete frequency, as "121.9," but more likely will shorten the instruction to "point niner.") When clear of the active runway—and past the hold line—come to a full stop and change to the Ground frequency:

You: Lincoln Ground, Cherokee One Four Six One Tango is clear of Runway Three-Five (or "the active"). Request taxi to the terminal. [Or you can request progressive taxi instructions as above.]

GC: *Cherokee Six One Tango, taxi to the terminal.*

You: Roger. Cherokee Six One Tango.

Remember to stay tuned to the Ground frequency any time you're moving the aircraft in a controlled area. Yes, you've landed and you've been cleared to taxi to the terminal, but just as in the taxiing-out example, Ground might need to give you subsequent instructions. This admonition is probably superfluous, but pilots have been known to switch everything off—rotating beacon, radios, transponder, and so forth—once cleared to the parking area. Except for the transponder, that's a no-no to those responsible for the control of airport ground traffic.

Moving the Aircraft from One Ground Location to Another (Required, Except on Ramp/Uncontrolled Areas)

Let's say you need to taxi to the other side of the field for a minor repair at a radio shop. To do so, you have to use a taxiway and cross the active runway. Permission for both is mandatory:

You: Lincoln Ground, Cherokee One Four Six One Tango at the terminal. Request taxi to King Radio.

GC: *Cherokee Six One Tango, taxi on Foxtrot. Hold short of Runway One Eight.*

You: Roger, hold short of Runway One Eight. Cherokee Six One Tango.

GC: *Cherokee Six One Tango, cleared to cross Runway One Eight.*

You: Roger. Cleared across One Eight. Cherokee Six One Tango.

Getting a Current Altimeter Setting (Optional)

You're ready to taxi from the ramp. After listening to the ATIS, however, you find that the altimeter setting given in the recording doesn't coincide with the field elevation. Realizing that the ATIS might be almost an hour old, you want the current reading. Incorporate your request for this data in the initial call to Ground:

You: Lincoln Ground, Cherokee One Four Six One Tango at the terminal with Information Lima. Ready to taxi. VFR Omaha. Request current altimeter setting.

GC: *Cherokee One Four Six One Tango, Lincoln altimeter is three zero zero five. Taxi to Runway Three Five.*

You: Roger, three zero zero five. Taxi to Three Five. Cherokee Six One Tango.

For a Radio Check (Optional)

You're on the ramp and want to determine the clarity and volume of your transmission. The easy way to do this is to call Ground and ask for a radio check:

You: Lincoln Ground, Cherokee One Four Six One Tango radio check.

GC: *Cherokee Six One Tango, loud and clear.*

You: Roger, Cherokee Six One Tango.

If Ground comes back with "transmission is scratchy and volume weak," don't indulge in an explanatory harangue. You've got a sick radio, Ground has told you so, so get off the air, and have a technician look at it. Ground can't help you a bit!

IS THE AIR CLEAR?

Now is as good a time as any to stress the point of listening before you speak. In other words, is the air clear?

You're at the ramp and have just tuned to Ground for taxi permission. The first thing that you hear is an IFR aircraft requesting its clearance. Ground responds with:

GC: *Cessna Three Four Romeo is cleared as filed to Denver. Maintain three thousand, expect ten thousand fifteen minutes after departure. Departure Control one two six point six. Squawk zero three two five. Fly heading two four zero after departure.*

[A brief period of silence]

34R: Cessna Three Four Romeo cleared as filed. Three thousand, ten thousand in fifteen. Departure one two six point six, zero three two five, two four zero heading.

GC: *Cessna Three Four Romeo, readback correct. Taxi to Runway One Eight.*

34R: Roger, taxi to One Eight. Cessna Three Four Romeo.

As indicated in brackets, there will almost always be a pause, a brief period of silence after Ground has conveyed the clearance, while the pilot is copying the instructions. Then comes the readback.

Don't start transmitting just because the air is momentarily quiet. Recognize what's taking place and give Three Four Romeo time to complete its task and reestablish communications with the controller. This courtesy applies to all communications situations, short of an emergency. Give both parties the opportunity to acknowledge instructions, answer a question, repeat an instruction, and the like. Momentary silence doesn't necessarily mean the air is all yours.

Pilots who don't listen and aren't considerate are usually the reason for the squeals and squeaks that distort reception. Two people can't transmit at the same time on the same frequency without creating that discordant cacophony that penetrates the cockpit or headset.

CONCLUSION

Ground Control is the "policeman" for all ground traffic—cars, trucks, tugs, and aircraft. Its use is mandatory. Even more than that, however, it is a source of assistance and an overseer of safety, ensuring the smooth flow of ground operations. Additionally, its very existence enables the other tower personnel who are responsible for the smooth flow of flight operations to concentrate solely on that full-time task.

It thus behooves all pilots to be familiar with what Ground Control can do for them. You must use the service, but use it wisely by being clear and concise and observing the basic rules of courtesy. The Ground Controller will invariably respond in kind.

10
Transponders

WHILE I HAVE REFERRED TO TRANSPONDERS IN SOME OF THE previous discussions and examples, this piece of electronic hardware hasn't been a major factor in the communication processes up to now. Henceforth, however, as I go through the sections on tower, Approach and Departure Control, and the Air Route Traffic Control Centers, the importance of the transponder increases—as does the importance of the terminology associated with it. Consequently, if you own or rent a transponder-equipped aircraft, familiarity with it is in order.

THE AIR TRAFFIC CONTROL RADAR BEACON SYSTEM (ATCRBS)

Simply said, the basic radar system is composed of two elements. One is the primary radar, which scans the surrounding area and produces images on the radarscope, or screen, of objects such as buildings, radio towers, aircraft without transponders, and aircraft with transponders turned off.

As this relatively minimal identification has safety and traffic control limitations, a secondary radar system was developed which incorporates a ground-based transmitter-receiver called an *interrogator*. This system—the Air Traffic Control Radar Beacon System (ATCRBS)—functions in unison with the primary radar and, in the scanning process, "interrogates" each operating transponder. In effect, it "asks" the transponder to reply. The primary and secondary signals are then synchronized and together transmit a distinctly shaped blip or target to the controller's radar screen. That blip, however, only tells the controller that there's an aircraft out there with a transponder "squawking" or "replying" with the standard VFR code of 1200. It does not permit more specific identification of the aircraft—which could be important in periods of heavy traffic or poor visibility.

Consequently, each transponder is equipped with an identification feature: the IDENT button. When the button is pushed, the radar target blip changes shape to distinguish the identing aircraft from other aircraft on the controller's screen.

Very broadly and nontechnically, this is the radar beacon system. For those interested in more detail, visit a radar-equipped tower or an Air Route Traffic Control Center. The specialists in either location are always glad to explain the system and let you watch it in operation—their workload permitting, of course.

THE TRANSPONDER: TYPES (OR MODES)

You'll frequently hear references to transponder "types" or "modes," and since several different modes exist, let's take a moment to clarify which is which:

- *Mode 1 and Mode 2* are assigned to the military
- *Mode 3/A* is used by both civilian and military aircraft
- *Mode B* is reserved for use in foreign countries
- *Mode C* is a Mode 3/A transponder equipped with altitude-reporting capabilities
- *Mode D* is not currently in use
- *Mode S* (the "S" stands for "Selective" address).

The last mode was intended to be the transponder of the future, what with its ability to function in concert with the FAA's TCAS (Traffic Alert and Collision Avoidance System); its built-in chip that can automatically transmit the aircraft's N-number, type, and altitude; and ATC's ability to selectively address to an individual aircraft inflight printouts of clearances, weather charts, and various other ground-to-air data. Whatever its fate down the road, the entire Mode S program is pretty much at a standstill in the late 1990s. And should it eventually become a full-blown reality, its costs will most likely make it an impractical investment for those who own or fly the typical general aviation light plane.

The transponder illustrated in Figs. 10-1 and 10-2, reproduced here courtesy of AlliedSignal Electronics and Avionics Systems, is a Bendix-King KT 76A. Along with the familiar function selector switch positions (keyed as no. 1 in Fig. 10-2), of OFF, SBY (standby), ON, ALT (altitude), and TST (test), the unit has four control knobs (no. 2, Fig. 10-2) with which the pilot can dial in whatever four-digit code Air Traffic Control assigns. Those numbers then appear in the code windows, no. 3, as illustrated in Figs. 10-1 and 10-2. Each of the four knobs can bring up numbers from 0 through 7, thus making possible 4,096 separate, or discrete, codes—($8 \times 8 \times 8 \times 8 = 4,096$). Hence the not-uncommon reference to a "4096 transponder."

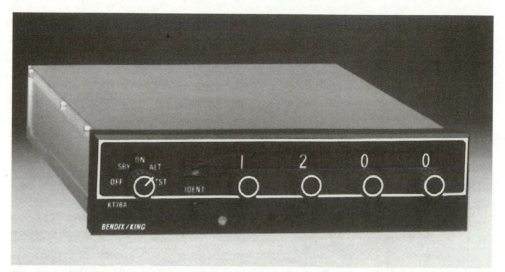

Fig. 10-1. *The shape and design of this Bendix-King transponder are typical of all Mode 3/A or Mode C transponders.*

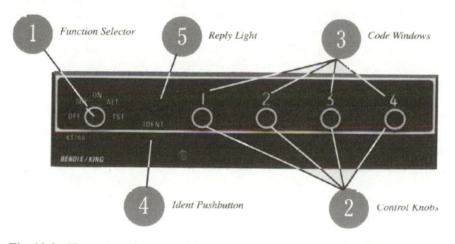

Fig. 10-2. *The various features of the KT76A transponder are identified in this AlliedSignal illustration.*

The Ident pushbutton (no. 4 in Fig. 10-2 but not really visible in the illustration) comes into play when ATC asks you to "Ident." In response, you push the button only once and the image on the controller's radar screen changes shape, thus permitting the controller to spot, or "identify," your particular aircraft more readily among the many that might be in the area or the traffic pattern. The thing to remember, though (and I'll discuss this more later), is that you never push that button until told by ATC to do so.

Finally, the transponder is equipped with a reply light (no. 5 in Fig. 10-2, again barely distinguishable here) which blinks every time the transponder responds to the radar beacon interrogator. These blinks also confirm to the pilot that the transponder is functioning. Just a point to be aware of: In some transponder makes, such as the Narco AT-50, the Ident button is incorporated in the reply light. Consequently, don't become confused if you happen upon a transponder that appears to be lacking the Ident feature. It's not missing; it's just part of the reply light "system."

If you're asked to "ident," and if you're one of only a few aircraft in the same general area, the controller can track you fairly easily once you and the others have "idented." It could be more difficult for him, though, if the activity is heavy, as around a busy airport. In these cases, he might ask you to report when over a certain landmark for verification of your position.

What I've said here describes the basic Mode 3/A unit. What Mode 3/A doesn't have is the altitude-reporting capability of Mode C. The 3/A is converted to Mode C merely by adding to the system either an *encoding altimeter* or a *blind encoder*. The first is a normal-appearing altimeter that is coupled to the transponder. As the altimeter aneroid bellows expand and contract with pressure changes, these changes are converted to coded response pulses by the transponder, which then transmits the aircraft's altitude (to the nearest 100 feet). The blind encoder performs the same function, but the unit is usually located behind the instrument panel, out of the pilot's sight.

Of the two, the blind encoder has certain advantages over the panel-mounted encoding altimeter. For one, it's usually a little less expensive and can be installed at about the same cost. Also, if it fails, it can be removed for repair and the aircraft operated as usual, except where Mode C is required.

On the other hand, if the encoding altimeter goes out and you still want or need altitude-reporting capability, the entire altimeter has to be removed and the aircraft is grounded pending repairs. Finally, installing an encoding altimeter means the replacement of what is probably a perfectly functioning unit. If you can sell the unit, fine. Otherwise, you've got a good altimeter to put on your fireplace mantle or office desk to impress visitors.

TRANSPONDERS: WHERE AND WHEN REQUIRED

With the concern for inflight safety heightened by recent midair collisions and reported near-misses, more stringent operating and equipment requirements have been established: Barring certain exceptions listed at the end of this section, it's almost mandatory today that every aircraft be equipped with an operable transponder. Otherwise, as the following indicates, freedom of flight is severely limited.

- The transponder *must* be in the ON position, or ALT position, if Mode C equipped, whenever you are operating in controlled airspace (which is just about everywhere in the United States).

- Mode C is required at and above 10,000 feet msl up to and within the Class A airspace, which starts at 18,000 feet msl. The only exception to the first portion of this particular rule is in certain mountainous regions in the western United

States where ground elevation rises to well above 10,000 feet msl. If you were flying at or below 2,500 feet aboveground in one of those areas, Mode C would not be required.

- A Mode C transponder is also required within 30 nautical miles of the Class B airports (currently the 33 largest and busiest in the country) from the surface up to 10,000 feet msl. This regulation involves the "Mode C veil," which can be identified on the sectional chart by the outermost thin blue circle that surrounds the entire Class B structure.

- Finally, the Mode C transponder is required in Class C airport areas from the surface up to 10,000 feet msl.

Exceptions to these transponder regulations do exist. For example, aircraft that were not originally certificated with an engine-driven electrical system are exempt, as are balloons and gliders *as long as* they avoid any Class B, C, or D airspace and remain below 10,000 feet msl. And then there are the exemptions granted to aircraft that operate into or out of certain airports located just inside that thin blue circle surrounding the airspace. The circle lies about 30 nautical miles out from the primary airport and establishes the outer limits of the Mode C veil itself.

This veil needs more explanation. For now, though, let's just acknowledge its existence and, barring certain exceptions, that it excludes all non–Mode C aircraft in the Class B airspace from the surface to 10,000 feet msl. Then, in Chap. 12, I'll review the structure of that airspace and its Mode C veil in more detail.

These various regulations do limit the airspace for non–Mode C aircraft, especially near the high density traffic centers, but that altitude-reporting capability has a lot of advantages, not the least of which is the added margin of safety it provides. Plus, it should keep pilots more on their toes and alert to what they're doing.

For example, if you accidentally or intentionally penetrate a Class B without approval, you can be sure that someone on the ground knows it. Your position, combined with your altitude readout from the Mode C, gives you away. Oh, you might be able to outrun the radar coverage and get home unidentified, but the FAA is continually developing more sophisticated means of tracking airspace violators. And once caught, it could mean a license suspension of 60 days or longer, plus a black mark on your record. With Mode C in operation, though, you know that you're probably being monitored. That ought to have a powerful effect on regulation abidance and attention to what you're doing.

TRANSPONDER OPERATION AND CODES

Operating the transponder is simply a matter of positioning the activating switch, being aware of the use of the Ident button, and, as I'll discuss shortly, entering the codes. A couple of hints about the operation, however, might be in order:

STANDBY: After engine start-up, turn on the radio(s) and put the transponder switch to the SBY position to allow the set to warm up without replying to the radar interrogator. Keep the switch in this position until cleared for takeoff.

ON or ALT: With Mode 3/A, switch from SBY to the ON (or in some units, NORMAL) position when cleared for takeoff, and leave it there throughout the flight, unless directed otherwise by ATC. If the unit has Mode C, switch to the ALT position. This not only turns the set on, but now it also reports your altitude. As mentioned earlier, when operating in controlled airspace (unless ATC directs otherwise), the transponder must be switched to ON (Mode 3/A), or, if equipped with an altitude encoder, ALT (Mode C).

OFF: Turn the transponder off as soon as you have landed, either on the rollout or when you're clear of the runway. If you leave it on, all you're doing is painting an enhanced image on the screen because of your proximity to the radar beacon antenna.

IDENT: Only when ATC asks you to "Ident" do you push the IDENT button—and just once. The signal sent will change the shape of the blip on the screen. Now the controller can more readily identify your aircraft and its relation to ground obstacles and other airborne traffic.

Above and beyond the basic operation of the transponder, it's essential that pilots be familiar with the standard numerical codes that are controlled by the four knobs on the set. The most common code is 1-2-0-0, which is the standard for all VFR altitudes and operations. There are exceptions to that statement, though, because if you are in contact with a Center or Approach or Departure Control, the controller will tell you to enter some other code (called a *discrete code*) for his specific identification of your aircraft.

Let's amplify that a bit. You've started out on a cross-country with the 1200 VFR code in the transponder. Then you call a Center and ask for "enroute traffic advisories." As a lot of VFR aircraft could be out there, the controller wants to spot your particular aircraft, so—and only for example—he asks you to "squawk 2056." This, or whatever four-digit combination he gives you, is a discrete computer-generated code that is assigned only to you. Once you have entered the new code, the computer recognizes it and displays your aircraft with a VFR symbol, your N-number, ground speed and, if Mode C-equipped, your altitude on the controller's screen. Now the controller has you distinguished from all other aircraft and is better able to alert you to possible conflicting traffic.

Other Codes

Along with 1-2-0-0, a few other standard codes have been established, as summarized in Table 10-1. Note, though, that the codes with asterisks are never to be used by civilian pilots.

TERMINOLOGY

Next, the terminology, or phraseology, associated with Mode 3/A and Mode C transponder operation:

Squawk: The origin of this rather odd term goes back to World War II and a radar beacon system called IFF (Identification, Friend or Foe). Allied aircraft were equipped with transmitters that replied to radar sweeps with a sound similar to a parrot's squawk. Today the term is used by controllers and pilots alike to indicate that the

Table 10-1. Transponder Codes

Code	Type of flight	When used
0000*	Military only	North American Air Defense
1200	VFR	All altitudes unless told otherwise
4000*	Military VFR/IFR	Warning/Restricted Areas
7500	VFR/IFR	Hijack
7600	VFR/IFR	Loss of radio communications
7700	VFR/IFR	Emergency ("Mayday")
7777*	Military only	Interceptor operations
Any other code assigned by ATC	VFR/IFR	When using Center or Approach/Departure Control

transponder is on and that a certain four-digit code should be, or has been, dialed in. If a controller asks you to "squawk two zero five six" (or any code), he wants you to enter those digits in the transponder. That is not an instruction to "ident," however. If you're squawking a code other than 1200 and are later told to "squawk VFR," it means to change the code back to 1200.

Ident: When a controller says, "Cherokee Six One Tango, ident," he or she wants you to push the little IDENT button so that he can more positively identify your aircraft. But push the button only once and only momentarily. Usually, but not always, after the radar target changes shape, the controller will come back to you with, "Cherokee Six One Tango, radar contact," or "Ident received." Then it's appropriate to acknowledge with, "Roger, Cherokee Six One Tango." On the other hand, when asked to ident, you don't have to reply with, "Roger, Cherokee Six One Tango identing." Just push the ident feature and say nothing. The blip change on the screen is acknowledgment enough for most controllers.

Occasionally, ATC may come back with, "Cherokee Six One Tango, I did not receive your ident. Ident again." In this case, acknowledgment is in order: "Roger, Cherokee Six One Tango." Now repeat the ident.

Squawk (number) and Ident: This instruction asks you to insert a certain numerical code and push the IDENT button. The word "and" might or might not be included in the instruction.

Stop Squawk: When you hear this, it means that the controller wants you to turn the transponder to OFF.

Squawk Standby: This instruction tells you to turn the switch from ON or ALT to SBY, the standby position. Remember, the transponder is not off. It's still warm but isn't responding to any interrogation and thus not transmitting. Your aircraft will be reflected on the screen only by the primary radar—not the secondary. (You might recall the pilot in Cherokee 41966 back in Chap. 1. He switched the transponder to STANDBY, not understanding that the controller wanted him to squawk 0252 and then standby for further instructions.)

Stop Altitude Squawk: If you have Mode C and hear this, merely switch from the ALT to the ON position. You're now functioning in Mode 3/A only, which provides aircraft identification but not altitude.

Squawk Mayday: You've verbally communicated an emergency to ATC, and for positive identification, the controller wants you to change your code to 7700.

THE TRANSPONDER AND RADIO COMMUNICATIONS IN AN EMERGENCY

You're cruising along when suddenly—or gradually—things begin to go wrong. Maybe it's a coughing engine, a dead engine, a fire, pilot illness, low fuel, you're lost, or whatever. In any event, action is necessary now, either to combat the immediate emergency or to prevent a potentially serious situation from becoming really serious. So, in the context of this subject, what should you do?

First, "Emergency" Definitions

Two terms are used to describe the nature of an emergency: One is *distress,* while the other is *urgency.* A *distress* condition is one involving a fire, a mechanical failure, or a structural failure. In other words, it's a condition that demands immediate attention and immediate assistance. An *urgency* is not necessarily perilous at the moment but could become potentially distressful. Examples: a low fuel supply, you're lost, heavy icing, pilot illness, weather, an unnatural engine vibration, or the like. In effect, an urgency warrants ground assistance, but the situation has not yet reached distress proportions.

Transponder Operation

When a distress or urgency situation arises, you should immediately enter the 7700 emergency squawk and then attempt to establish radio communications with an ATC facility.

The 7700 code appears on the screens of all radar-equipped facilities within range of their radar coverage, and, by sound as well as a flashing blip on the screen, attracts the controller's attention. Even if not in radio contact with the aircraft, the facility, or facilities, is alerted to the existence of an aircraft out there in trouble.

Radio Communications

If you've been in contact with a tower, a Control Center, or an FSS when either a distress or urgency develops, the communications should obviously be directed to that facility. A distress message should start with the word "Mayday" and be repeated at least three times. ("Mayday" comes from the French, "M'aidez," meaning "Help me.") In an urgency, "Pan-pan," again repeated three times, starts the radio transmission.

Distress messages have absolute priority over all others, and the word "Mayday" commands radio silence on the frequency in use. Urgency messages have priority over all others except distress. "Pan-pan" thus warns other stations not to interfere with urgency transmissions. That noninterference applies to you as well. If you hear either "Mayday" or "Pan-pan," stay off the air so that the ground facility can have uninterrupted communications with the troubled aircraft.

If you have not been receiving services from a facility, such as a Center or a tower, and, if time permits, the Center, tower, or FSS in whose area of responsibility you're flying should be contacted on the appropriate frequency. By so doing, a more rapid response to the call is likely—a good reason for listing the various enroute frequencies in the preflight planning and having the list immediately available in the cockpit.

Conditions might be such, though, that you have no time to search for the correct frequency, even with a prepared list. In such cases, two other frequencies can be used: the *emergency-only* frequencies of 121.5 or 243.0. Both have ranges generally limited to line of sight. The frequency 121.5 is guarded by direction-finding stations and some military and civil aircraft. Frequency 243.0 is guarded by military aircraft. In addition, both frequencies are guarded by military towers, most civil towers, FSSs, and radar facilities.

The logical facility to try to reach first on 121.5 is the nearest ARTCC. The ARTCC's emergency frequency capability, however, does not normally extend to its radar coverage limits, so if you get no response, address the call to the nearest tower or FSS or to "any station...."

Recognizing the problem of time in a distress condition, as much of the following as possible, preferably in the order suggested, should be communicated:

1. "Mayday, Mayday, Mayday" or "Pan-Pan, Pan-Pan, Pan-Pan"
2. Name of facility addressed or "any station"
3. Aircraft identification and type
4. Nature of distress or urgency
5. Weather
6. Pilot's intentions and request
7. Present position and heading, or if lost, last known position, time, and heading since that position
8. Altitude or flight level
9. Hours and minutes of fuel remaining
10. Any other useful information, such as visible landmarks, aircraft color, emergency equipment on board, number of people on board
11. Activate the Emergency Locator Transmitter (ELT), if installation permits

As an illustration of a distress call: Assume that you're about 50 miles east of Atlanta on Victor Airway 18, destination Columbia, South Carolina. Along the way, you've been monitoring Atlanta Center to get an idea of the traffic in the area, but you haven't been in contact with the facility for traffic advisories. Suddenly you have an oil line break. The pressure gauge drops rapidly and the spurting oil covers the windshield with a coat of film that reduces forward visibility almost to zero. You've got a problem.

The first thing to do is enter 7700 in the transponder. Then get on the air to Atlanta Center—Atlanta because the frequency is already tuned in and no time is wasted changing to 121.5. Even though you haven't been talking to the Center, the fact that you're listening on the frequency means that you can establish immediate contact with the controller responsible for your sector or area.

Now an example of the call—a call that is as concise as possible, communicating only the most essential information.

> **You:** Mayday, Mayday, Mayday. Atlanta, Cherokee One Four Six One Tango distress. Oil line break. No forward visibility account break. Land immediately. Fifty east, Victor 18 at six point three [that's pilot jargon for "six thousand three hundred" feet altitude], descending. White, red stripes. Two aboard.

The minimum initial information the controller needs is (1) aircraft type and N-number, (2) the problem that is creating the emergency, and (3) intentions. Then, if conditions permit, volunteer the rest of the information listed in the example.

Note that items 5 and 9 of the information listed above have not been addressed in this example. They're omitted because the weather is presumably VFR or you wouldn't be there in the first place, and fuel remaining is hardly consequential in a mechanical emergency. If you were lost or low on fuel, both elements would be important to people on the ground, but not here.

Note, too, that there's no unnecessary verbiage. Speak as calmly and clearly as possible, but make the messages short, terse, to the point.

I won't simulate further exchanges, because the main burden for instructions now falls on the controller. Your job is (1) to maintain control of the airplane; (2) to comply, to the best of your ability, with his instructions; and (3) to keep the controller informed of what's taking place.

Pilot Responsibilities after Establishing Radio Contact

Once you are in contact with a ground facility and have conveyed as much of the above information as time and pertinence permits, your responsibilities from this point are these:

- Comply with advice and instructions, if at all possible.
- Cooperate.
- Ask questions or clarify instructions not understood or with which you cannot comply.
- Assist the ground facility in controlling communications on the frequency. Silence interfering stations.
- Do not change frequencies or change to another ground facility unless absolutely necessary.

- If you do change frequencies, always advise the ground facility of the new frequency and station before making the change.

- If two-way communication with the new frequency cannot be established, return immediately to the frequency where communication last existed.

- Remember the emergency four Cs:

  Confess the predicament to any ground station;
  Communicate as much of the distress/urgency message as possible;
  Comply with instructions and advice;
  Climb, if possible, for better radar detection and radio contact.

EMERGENCY LOCATOR TRANSMITTER (ELT)

While not associated with transponder operations, the Emergency Locator Transmitter (ELT) is related to the subject at hand and is an important element in emergency search-and-rescue (SAR) efforts. In essence, the ELT exists solely to assist in locating downed aircraft.

ELTs are required equipment for most general aviation aircraft, per FAR Part 91.207, although that part also lists certain exceptions. Essentially, an ELT is nothing but a battery-operated transmitter, along with an externally mounted antenna, that transmits a continuous and distinctive audio signal on the 121.5 and 243.0 frequencies when subjected to crash-generated forces. When transmitting, the life of an ELT is 48 hours over a wide range of temperatures.

Depending on its location in the aircraft, some ELTs can be activated by the pilot in flight. In other installations, the ELT is secured elsewhere inside the fuselage and can't be accessed except on the ground. With this remoted installation (usually near the tail of the aircraft), the transmitter is activated by ground impact, or the pilot can do so after a survivable forced landing when ground impact is comparable to a normal landing.

Because of their importance in SAR efforts, ELT batteries are legal for only 50 percent of their manufacturer-established shelf lives, after which they must be replaced. In the interim, periodic ground checks should be made to determine the battery's viability. These checks, though, can be made only on the ground and in accordance with FAR regulations or the "Emergency Procedures" chapter in *AIM*.

One more note: The FAA urges all pilots to monitor 121.5 while flying around. You might pick up an ELT signal that was beyond the range of an FSS or ATC facility and possibly, through radio contact with a nearby facility, assist in the SAR operation. So, become familiar with what an ELT transmission sounds like and be prepared, if you can, to help a fellow pilot in distress.

CONCLUSION

I've gone into the matter of emergencies in some detail here because of the roles the transponder and radio transmissions play when trouble brews aloft. As rare as troubles are, especially those of a mechanical nature, each of us who flies should be prepared for the worst. That's one reason simulated forced landings, stalls, short and soft field landings, and so on, are so much a part of initial and recurrent training.

A valuable question to include in any preflight planning is, "What could go wrong?" Then follow that question with another: "If what could go wrong did go wrong, what will I do?" In business parlance, it's called "PPA"—potential problem analysis. Not only is the "what could go wrong" analyzed, but so are the possible solutions or plans of action if the potential became reality. Those who know what to do, who have planned and practiced, are those who have conditioned themselves to confront airborne problems with reasonable calmness and confidence. Thorough preparation is about 90 percent of emergency survival.

A FEW REMINDERS AND TIPS

As a brief summary, a few reminders and tips about the transponder are perhaps in order:

- Remember to put the transponder in the SBY position after engine start. Change it to ON or ALT only after being cleared for takeoff or during the takeoff roll. Leave it on throughout the flight (mandatory), unless directed otherwise by an ATC facility. On landing, turn it OFF either on touchdown or when you've cleared the runway. Don't keep it in the ON or ALT position while you're taxiing in or out—whether the airport is controlled or not.

- When told by a controller to change from one code to another, jot down the new code on a piece of knee pad paper and then repeat the code back to the controller:

ATC: *Cherokee One Four Six One Tango, squawk five three four zero and ident.*

You: Roger, five three four zero. Six One Tango.

Now push the "Ident" button.

The reason for the readback: It's easy to confuse or transpose digits, such as 5-4-3-0 with 4-5-3-0, or any other combination. The readback catches any discrepancy. Also, writing down the new code will help you get it straight. It's kind of embarrassing to have to go back to the controller and sheepishly ask, "What was that squawk you gave me?"

- If you have a transponder, with or without Mode C, make it known to ATC in your initial contact:

Turner Tower, Cherokee One Four Six One Tango over Cordele at three thousand five hundred landing Albany, squawking one two zero zero with Information Delta. Nowadays, controllers shouldn't have to ask, "Are you transponder equipped?" but they still do, and you still hear it.

- When changing codes, always avoid even a momentary display of 7500, 7600, or 7700. Cycle your changes so that those emergency or radio-loss codes never appear during normal flight operations.

- Become familiar with the transponder radio terminology, as "Squawk VFR," "Squawk Standby," and so on. Again, it's embarrassing to have to ask the controller what he wants you to do.

• Know the emergency transponder and radio procedures by heart. You'll probably never need them, but, like a good Scout, "Be prepared."

SUMMARY

The transponder is an important element in air traffic control and the ever-increasing drive for safety, and becomes even more essential as new FAA regulations go into effect.

The nice thing about a transponder, though, is that it doesn't require a lot of pilot expertise. In some respects, it's a little like ATIS: It does more for you than you have to do yourself. With ATIS, you just sit and listen. With the transponder, you turn it on, understand the limited phraseology, do as you're asked, and that's it. It—not you—keeps the people on the ground informed of where you are, and, if Mode C–equipped, your altitude. Functioning as it does, it reduces the need for voice communications and contributes to the safety of all of us.

But, a final word is essential. The transponder and what it does should never lull you into complacency. Despite the controller's skill and the sophistication of his electronic equipment, it's still the pilot's job to see and avoid. IFR aircraft in instrument meteorological conditions within controlled airspace are assured of horizontal and vertical separation from all other aircraft. As a VFR pilot, however, the most you'll usually get are advisories of the positions of other aircraft and safety alerts when known conflicts seem imminent. In a Class B the control is greater. Otherwise, it's up to you, with whatever help ATC can give you, to do what you always should be doing—protecting your own skin. The final responsibility sits in the left seat of every airplane.

11
Operating and Communicating in Class D and E Airspaces

GROUND CONTROL HAS CLEARED YOU TO TAXI TO (NOT ON) THE active runway. You've completed the pretakeoff check and have moved from the runup area to the runway hold line. With the aircraft at a full stop, you change to the tower frequency to request takeoff clearance....

About to come into play is the simplest element of the *controlled* airspace system: the Class D airport, its surrounding airspace, and what are called *transition areas*—areas designed to facilitate the arrival and departure of IFR aircraft in instrument meteorological conditions (IMC). Consequently, to ensure understanding of these various airspace components, let's start with a brief overview of the Control Tower, followed by a discussion of the Class D airspace structure itself and the associated transition areas. Almost simultaneously, the Class E airspace comes into play because of the fact that when a

Class D tower closes down, as many do during the nighttime hours, that airspace becomes either a Class E or a Class G—depending on the availability of a weather reporting source. Following those general airspace reviews, we'll get to specific examples of FAA-approved radio procedures when communicating in this Class D airspace. So with that, this question arises…

WHAT DOES THE CONTROL TOWER DO?

Perhaps the question is so basic that the answers are obvious. On the other hand, perhaps not. Airspace violations and messed-up radio talk raise doubts about how well some pilots understand the tower's responsibilities. So does the reluctance of many pilots to venture into a tower-controlled airport if they've been trained at small uncontrolled fields. Anyway, necessary or not, let's pursue the subject.

Simply said, the tower controls all traffic in the Class D airspace, including ground movements on the airport, as cited in Chap. 9. In other words:

- To land at or take off from a tower-controlled airport, you must first establish radio contact with the tower controller, obtain specific clearance to take off or land, and then maintain radio contact while within the Class D airspace.

- To fly through any portion of the Class D airspace, you must advise the tower controller of your intentions, obtain clearance to enter the airspace, and thereafter follow instructions or vectoring.

- If the controller gives you instructions that would cause you to violate a VFR regulation or place you in jeopardy, you must advise him or her of the potential problem and the evasive actions you are taking, or intend to take.

Just remember that the tower is not an advisory service, as is unicom or a Flight Service Station. It's a *controlling facility*, which means that its directions must be followed, except in a bona fide emergency or in the situation cited immediately above.

In effect, the controller in the tower is the airport policeman. It is his or her duty to maintain order in the airspace and to ensure the safe, efficient flow of traffic into, out of, and within that airspace.

THE CLASS D DIMENSIONS AND KEY FEATURES

The Class D airspace can be easily spotted on the sectional by its blue airport symbol. Whenever you see that symbol, you'll know that the airport has an operating tower and that the airport is thus controlled. That's not enough, though, because Class C, Class B, and TRSA airports are also depicted in the same blue color, although each also has its own distinctive surrounding circular rings or designs and color codings that I'll describe more fully later. Thus for Class D airspace, the other identifying feature is the blue segmented circle that surrounds the primary airport. This circle (it's *almost* always a circle), which has a radius of 4.3 nautical miles from the center of the airport (5.0 statute miles), defines the horizontal limits of the airspace. Class B, C, and TRSA airport towers have the same approximate radius of control, but those limits are not so identified on the chart.

Keep this in mind, however: If the Class D tower is in use part-time, the airport is classified as uncontrolled when the tower is closed and the airspace becomes Class E. That is, it becomes Class E *if* a weather-reporting service, human or automated, is available on the airport. Lacking any sort of weather service, the airspace below 700 feet reverts to a Class G.

Let's be more specific about that and the requirements to justify designating an airport a Class D or E. Basically, *at least one* of the following that can provide the essential weather service must be located on the airport:

- An operating tower and a qualified weather observer
- An operating Flight Service Station
- A National Weather Service office
- A qualified weather-observing source [human or ASOS/AWOS (Automated Surface Observation System/Automated Weather Observing System)] (more on this subject in a moment)

A Class E airport/airspace obviously doesn't have an operating control tower; if it did, it would be a Class D. Also, only a few airports these days have a Flight Service Station on site, and the likelihood of a National Weather office being on the property is even more remote. So, we are left with the only conclusion: A qualified weather-observing source must be available. Hence, the Class E airspace designation is used.

While the airport is in the Class E or G status, and assuming you're flying VFR in VFR weather, no radio contact with a controlling agency is required. In fact, none is available. Yes, that last statement is true even if a Flight Service Station is on the field. Remember, Flight Service Stations are not flight or traffic controlling agencies. Consequently, operations are exactly as at a unicom or multicom field. You can enter the area, land, take off, shoot touch-and-gos—whatever, without approval from anyone. All you need to do, radiowise, is make the calls illustrated back in those earlier chapters on multicom and unicom. Do keep this in mind, though: With a tower on the field, even if it's closed, all radio transmissions are made over the *tower's* CTAF frequency, not the unicom.

A BRIEF SUMMARY OF ASOS AND AWOS

Since the subject was just raised, a momentary interruption to summarize those automated reporting systems would seem to be in order. First off, ASOS is the primary airport surface reporting system which, according to *AIM,* "(the system) will (sic) provide continuous, minute-by-minute observations and perform the basic observing functions necessary to generate an aviation routine weather report (METAR) and other aviation weather information."

If located on a given airport, either "ASOS" or "AWOS," along with its discrete radio frequency, appears on the sectional chart adjacent to the airport information (Fig. 11-1). Similarly, the availability of ASOS or AWOS is noted in the *A/FD,* immediately following the section headed "Airport Remarks" (Fig. 11-2). In some cases, especially at the more active airports where weather information via sources such as ATIS, Approach Control, or tower personnel is available, ASOS, if it exists, can be accessed only by telephone.

Minute by minute, each ASOS gathers through its sensors and then reports airport cloud heights, visibility, precipitation, pressure, temperature, dew point, wind direction, and speed. The reporting is accomplished via computerized voice transmission over the published VHF frequency or the voice portion of a local navaid. Normal reception is within 25 miles of the airport, up to 10,000 feet agl.

AWOS is the other weather-observing system, but ASOS-3 incorporates all of the information in AWOS reports, plus precipitation identification and intensity. More extensive than AWOS, ASOS has become the logical observing-and-reporting system.

One problem with both systems, however, is that up until now, at least, they can only report conditions vertically—that is, from their individual positions on the ground straight up, not horizontally or on a rising slant. Therein lies the potential problem: The weather overhead is great, but just off the field in any direction could lurk a huge black cloud full of lightning, thunder, and all forms of precipitation. Except possibly for recording increased winds preceding the nearby storm, these less-than-desirable flying conditions would go unreported.

There's obviously more to the ASOS/AWOS story than this brief overview, so for those interested in additional information about the systems, check out Chap. 7 of *AIM*. The three or four pages there will fill in many of the blanks.

Fig. 11-1. *Among other information, the sectional identifies the existence of ASOS or AWOS on the airport.*

```
MANKATO MUNI    (MKT)  5 NE   UTC−6(−5DT)   N44°13.30' W93°55.13'                              OMAHA
   1020    B   S4   FUEL 100LL, JET A, MOGAS   OX 4   ARFF Index Ltd.                  H−1E, 3G, L−10G, 11C, A
  RWY 15−33: H5400X100 (ASPH−GRVD)     S−40, D−60, DT−100   HIRL                                   IAP
    RWY 15: REIL. PAPI(P4L)—GA 3.0° TCH 48'. Road.        RWY 33: MALSR. PAPI(P4L)—GA 3.0° TCH 47'.
  RWY 04−22: H3999X75 (ASPH−GRVD)     S−18, D−25   MIRL
    RWY 04: REIL. VASI(V2L)—GA 3.0° TCH 39'.        RWY 22: REIL. VASI(V2L)—GA 3.0° TCH 26'. Trees.
  AIRPORT REMARKS: Attended 1200−0500Z‡. CLOSED Christmas. Deer on rwys and in vicinity of arpt. Migratory
     waterfowl on and in vicinity of arpt. PPR 24 hours for air carrier ops with more than 30 passenger seats call arpt
     manager 507−625−6006. ACTIVATE MIRL Rwy 04−22, VASI and REIL Rwy 04, VASI and REIL Rwy 22, HIRL Rwy
     15−33, PAPI and REIL Rwy 15, PAPI and MALSR Rwy 33—CTAF.
→ WEATHER DATA SOURCES: AWOS−3 110.8 (507) 625−3726.
  COMMUNICATIONS: CTAF/UNICOM 122.7
    PRINCETON FSS (PNM) TF 1−800−WX−BRIEF. NOTAM FILE MKT.
    RCO 122.1R 110.8T (PRINCETON FSS)
  ® MINNEAPOLIS CENTER APP/DEP CON 135.0
  AIRSPACE: CLASS E svc Mon−Fri 1100−0530Z‡. Sat 1300−1600Z‡ and 1900−2200Z‡. Sun 2130−0530Z‡ other times
     CLASS G.
  RADIO AIDS TO NAVIGATION: NOTAM FILE MKT.
    (L) VOR/DME 110.8    MKT    Chan 45   N44°13.20' W93°54.74'    at fld. 1020/7E.
      DME portion unusable 325°−150° beyond 16 NM.
    ILS 108.7   I−MKT    Rwy 33.   Localizer unmonitored.
```

Fig. 11-2. Another source of ASOS or AWOS information, where the facility exists, is the A/FD.

OTHER FEATURES OF THE CLASS D AIRSPACE

• As I've already indicated, the tower exercises complete control over the flow of all traffic—on the ground at the primary airport and airborne within the limits of its airspace.

• Ceilings of a Class D airspace are generally 2,500 agl, but they do vary upward somewhat with individual airports. Above the ceiling, the pilot is in Class E airspace, unless the airport directly underlies a Class B or Class C airspace. If operating VFR in that E airspace, no radio contact with the airport tower is necessary or wanted.

• To enter a Class D airspace, whether landing at the primary airport, at another airport within that airspace, or just to fly through it to get to the other side, radio contact with the tower *must* be established before penetrating any portion of the airspace.

CLASSES D AND E TRANSITION AREAS

The sole purpose of transition areas in the Class D and E airports/airspaces is to provide proper separation of arriving and departing IFR aircraft during instrument meteorological conditions (IMC). Among other actions, this could mean prohibiting or minimizing VFR traffic within the airspace when conditions are below normal VFR minimums. [I'm referring here, of course, to Special VFR operations (SVFR) discussed in Chap. 7 on Flight Service Stations.]

A logical question at this point: If the airspace is Class D, what is the probability of it having radar on site and being capable of providing radar control and instructions—in other words, its own Approach/Departure Control facility? The answer is, and only a broad generality: about 50-50, maybe a little better (I haven't been able to get a definitive answer to that).

The next question then, is: How do you determine whether there is Approach Control on a given Class D airport? There's only one answer to that: You have to refer to the *Airport/Facility Directory*. There's nothing on the sectional that gives you any clue.

As one example, take Clarksburg's (West Virginia) Benedum Airport (Fig. 11-3). From all appearances, Benedum is just a standard tower-only Class D airport and airspace. If you wanted to land there, however, a glance at the Northeast U.S. *A/FD* (Fig. 11-4) would tell you that radar Approach/Departure Control is located on the field. The clue is the symbol ®. This *A/FD* excerpt also states that the facility is open from 1200 to 04000Z, the same hours as the tower. Furthermore, when the tower and Approach are closed, the Cleveland Air Route Traffic Control Center (ARTCC) assumes the approach/departure responsibility through its remote radar outlet. Philadelphia does the same for Wilmington, Delaware (Fig. 11-5); and so on. One way or another, most Class D and E airports, perhaps even hundreds of miles from a Control Center, have the assurance of radar coverage during IMC weather.

One other aspect of this: A few Class D towers, located near a Class B or C Approach Control facility, might be equipped with a receiver called BRITE (Bright Radar Tower Equipment). What happens in such cases is that Approach receives the radar images within its range of coverage and transmits a video compression of those same images by a

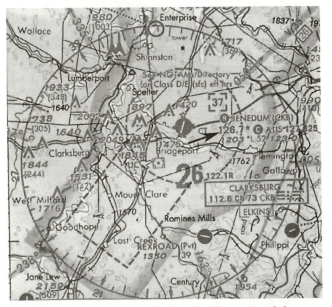

Fig. 11-3. One bit of information that the sectional does not indicate is the existence or frequency of Approach/Departure Control at a Class B, C, or D airport.

CLARKSBURG
    BENEDUM   (CKB)   0 NE   UTC–5(–4DT)   N39°17.66′ W80°13.76′   **CINCINNATI**
      1203   B   S2   **FUEL** 80, 100LL, JET A   OX 1, 2   ARFF Index A.   **H–4I, 6H, L–22F, 23D, 24E**
      RWY 03–21: H5198X150 (ASPH–GRVD)   S–70, D–90   HIRL   0.4% up NE   **IAP**
        RWY 03: REIL. VASI(V4L)—GA 3.44°TCH 59′. Trees.   **RWY 21:** MALSR. Trees.
      AIRPORT REMARKS: Attended 1000–0500Z‡. Fuel not avbl during hours 0500–1000Z‡. Deer on and in vicinity of arpt.
        PPR 24 hours for air carrier operations with more than 30 passenger seats call arpt manager 304–842–3400.
        ACTIVATE HIRL Rwy 03–21—CTAF. Ldg fee for all acft over 6500 lbs.
      WEATHER DATA SOURCES: LAWRS.
      COMMUNICATIONS: CTAF 126.7   **ATIS** 127.825(1200–0400Z‡)   **UNICOM** 123.0
        ELKINS FSS (EKN) TF 1–800–WX–BRIEF. NOTAM FILE CKB.
        CLARKSBURG RCO 122.1R 112.6T (ELKINS FSS)
→ Ⓡ **CLARKSBURG APP/DEP CON** 119.6 (West) 121.15 (East) (1200–0400Z‡)
→ Ⓡ **CLEVELAND CENTER APP/DEP CON** 126.95 (0400–1200Z‡)
        CLARKSBURG TOWER 126.7 (1200–0400Z‡)   **GND CON** 121.9
      AIRSPACE: CLASS D svc effective 1200–0400Z‡ other times CLASS G.
      RADIO AIDS TO NAVIGATION: NOTAM FILE CKB.
        CLARKSBURG (L) VOR/DME 112.6   CKB   Chan 73   N39°15.19′ W80°16.07′   040° 3.1 NM to fld. 1430/04W.
        ILS 109.3   I–CKB   Rwy 21.   LOC unusable byd 25° rgt of course. GS unusable blo 1600′. ILS unmonitored
          when twr clsd.
   • • • • • • • • • • • • • • • • • • •
      HELIPAD H1: H50X50 (CONC)
      HELIPAD H2: H50X50 (CONC)
      HELIPORT REMARKS: Helipad H1 located on FBO apron. Helipad H2 located on FBO apron.

Fig. 11-4. *The A/FD is the only source that provides Approach/Departure information.*

WILMINGTON
    NEW CASTLE CO   (ILG)   4 S   UTC–5(–4DT)   N39°40.72′ W75°36.39′   **WASHINGTON**
      80   B   S4   **FUEL** 100LL, JET A,   OX 1, 2, 3, 4   LRA   ARFF Index A   **H–6I, L–24G, 28F**
      RWY 09–27: H7165X150 (ASPH–GRVD)   S–90, D–140, DT–250   HIRL   **IAP**
        RWY 09: ODALS. PAPI(P4L)—GA 3.0° TCH 55′. Ground.   **RWY 27:** VASI(V4L)—GA 3.0°TCH 51′. Trees.
      RWY 01–19: H7002X200 (ASPH–GRVD)   S–90, D–140, DT–250   HIRL
        RWY 01: ASLF1. Road.   **RWY 19:** VASI(V4L)—GA 3.0°TCH 58′. Trees.
      RWY 14–32: H4596X150 (ASPH)   S–50, D–60   MIRL
        RWY 14: Trees.   **RWY 32:** VASI(V4L)—GA 3.0°TCH 27′.
      AIRPORT REMARKS: Attended continuously. Birds on and invof arpt. Rwy 09–27 no touch and go ldg for turbo jet
        0400–1200Z‡. Rwy 14–32 restricted to acft 60000 lbs or less. PPR one hour for unscheduled air carrier
        operations with more than 30 passenger seats call 302–571–2913 (digital pager). From 0300–1100Z‡
        announce emerg on CTAF direct to the ARFF station.
        Rwy 14–32 CLOSED to FAR Part 121 operators except 1 hr PPR call 302–571–2913 (digital pager). Extensive
        pilot training for fixed wing and helicopters. When terminal building clsd 0400–1100Z‡ contact arpt safety
        department on 121.7 or 302–571–2913 (digital pager). Twy G section between Twys H and J limited to acft with
        wingspan of 79 ft or less. Twy G section between Twys H and D limited to acft with wingspan of 49 ft or less.
        When twr clsd ACTIVATE HIRL Rwys 01–19; 09–27; twy lgts; ALSF1 Rwy 01 and ODALS Rwy 09—CTAF. MIRL Rwy
        14–32 and Twys 'B' 'C' and 'D' lgts preset on med ints. Ldg fee for all acft except federal government; military
        and based single engine acft. NOTE: See Land and Hold Short Operations Section.
      WEATHER DATA SOURCES: ASOS (302) 322-8451.
      COMMUNICATIONS: CTAF 126.0   **ATIS** 123.95 (1200–0400Z‡)   **UNICOM** 122.95
        MILLVILLE FSS (MIV) TF 1–800–WX–BRIEF. NOTAM FILE ILG.
        DUPONT RCO 122.1R 114.0T (MILLVILLE FSS)
→ Ⓡ **PHILADELPHIA APP/DEP CON** 118.35   **CLNC DEL** TF 800–354–9884
        WILMINGTON TOWER 126.0 (1200–0400Z‡)   **GND CON** 121.7
      AIRSPACE: CLASS D svc 1200–0400Z‡ other times CLASS E.
      RADIO AIDS TO NAVIGATION: NOTAM FILE ILG.
        DUPONT (L) VORTAC 114.0   DQO   Chan 87   N39°40.69′ W75°36.43′   at fld. 71/10W.
        HADIN NDB (LOM) 248   IL   N39°34.87′ W75°36.84′   013° 5.9 NM to fld.
        ILS 110.3   I–ILG   Rwy 01.   LOM HADIN NDB. ILS unmonitored when twr clsd.
   • • • • • • • • • • • • • • • • • •
      HELIPAD H1: H50X50 (ASPH)
      HELIPORT REMARKS: Heliport located NW of Twy 'A' opposite terminal building.

Fig. 11-5. *Philadelphia is responsible for approach and departure control at Wilmington's New Castle County Airport.*

phone line to the nearby Class D tower. These second-generation images then equip the D-tower controller to provide radar-influenced traffic advisories and instructions to transponder-equipped aircraft within the Class D airspace.

MORE SECTIONAL CHART SYMBOLS AND WHAT THEY MEAN

While on the subject of protecting IFR aircraft in IMC weather, you've undoubtedly noticed on the sectional chart the numerous circular as well as irregularly shaped magenta-colored designs surrounding so many airports, particularly the Class D, E, and G airports. These designs represent the areas where the floor of the Class E airspace drops to 700 feet agl as opposed to its normal 1,200-foot low. By so doing, that much more space is made available for controlling departing or arriving IFR traffic and, in the process, limiting or even eliminating VFR activity during IMC weather. (This is again a reference to those SVFRs.)

The outer edge of the symbol, you'll note, is sharp and clearly defined, while the coloring fades a little and becomes fuzzy on the inner side of the design. This fuzziness indicates that the 700-foot floor prevails everywhere within the symbol. Outside of the design or symbol, it's all Class E airspace until you run into a Class B, C, or D airspace, another airport with the same basic magenta design, or you climb until you reach the 18,000-foot floor of the Class A airspace.

Something else to note about the Class D and E airport areas: If you'll check the sectional, such as Fig. 11-6 and the Yuma, AZ, International Airport, you'll see the blue segmented circle (it's *usually* a circle) around the Class D airport. Also, although you can't determine it in this black-and-white figure, the two extensions at the 12 and 1 o'clock positions are in magenta. The extensions are added to the airspace, and thus to the design, to keep IFR operations within controlled airspace during IMC conditions. When any one extension reaches out more than 2 nautical miles from the normal circular airspace limit, (1) all extensions become Class E; (2) all extensions are colored magenta on the sectional; (3) VFR traffic within the extensions is subject to VFR weather minimums; and (4) VFR traffic does not require radio communications with the tower in normal VFR weather. In other words, just being attached to the circle and to the area in magenta indicates that that area is Class E and radio contact to enter same is not necessary.

For a different adaptation of the extensions, take a look at Fig. 11-7, New Mexico's Cannon Air Force Base. This is a Class D airport, but the small extensions that jut out from the circle (as opposed to being attached, as in the Yuma illustration) indicate that they are literally part of the Class D airspace. That being the case, contact with the control tower must be established before entering any portion of those extensions as well, of course, as the airspace within the rest of the circled area.

One other point about the the Cannon example: Note the "68" in the segmented square box above the airport symbol. This indicates that the ceiling of the Class D airport is 6800 feet above mean sea level, or 2505 feet above the airport itself, which, per the sectional chart, is 4295 msl.

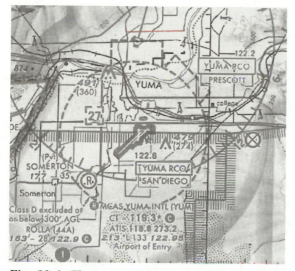

Fig. 11-6. *The segmented extensions at the 12:00 and 1:00 o'clock positions indicate a Class E–controlled airport in IMC (Instrument Meteorological Conditions).*

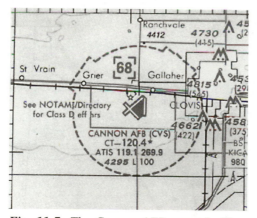

Fig. 11-7. *The Cannon AFB example illustrates how an extension of the Class C airspace is depicted on the sectional.*

DETERMINING THE TOWER FREQUENCY

For the VFR pilot, two sources are available to determine tower frequencies—the sectional chart and the *Airport/Facility Directory (A/FD).*

The sectional always publishes the frequencies in two locations: adjacent to the airport name, with "CT" preceding the frequency (Fig. 11-8); and on the reverse side (usually) of the legend flap (Fig. 11-9).

Fig. 11-8. *The sectional identifies the Elizabeth City Control Tower frequency as 120.5, both when the tower is open and when its CTAF is closed.*

CONTROL TOWER	OPERATES	TWR FREQ	GND CON	ATIS	ASR/PAR
ANDREWS AFB/NAF	CONTINUOUS	118.4 289.6	121.8 275.8	113.1 251.05	ASR
ATLANTIC CITY INTL	CONTINUOUS	120.3 239.0	121.9 284.6	108.6	ASR
BALTIMORE-WASHINGTON INTL	CONTINUOUS	119.4 257.8	121.9	115.1 127.8	
BLACKSTONE AAF PERKINSON	BY NOTAM	126.2 241.0			
CAPITAL CITY	0600-2200	119.5 257.8	121.9	134.95	
CHARLOTTESVILLE-ALBEMARLE	0700-2300	124.5 242.7	121.9 242.7	118.425	
DAVISON AAF	0600-2200 MON-FRI EXC HOL	126.3 229.4	121.9 245.2	128.175	ASR/PAR
DOVER AFB	CONTINUOUS	126.35 327.5	121.9 225.4	135.05 273.5	ASR
EASTERN WV REGIONAL/ SHEPHERD	0700-2200 TUE-THU 0700-1600 FRI-SAT 1300-1800 SUN O/T BY NOTAM	124.3 236.6	121.8 275.8		
ELIZABETH CITY CGAS	0700-2200	120.5 355.6	121.9		
FELKER AAF	0700-2300 MON-FRI EXC SAT-SUN & HOL	126.3 248.2	121.35 229.4		
HARRISBURG INTL	CONTINUOUS	124.8 231.1	121.7 348.6	118.8	
NAES LAKEHURST/ MAXFIELD	NOV-APR 0700-1900 MON-FRI MAY-OCT 0700-1900 MON, WED, FRI 0700-2100 TUE, THU CLSD SAT-SUN & FED HOL	127.775 360.2	121.0 352.4		
LANCASTER	0600-2300	120.9 251.1	121.8	125.675	
LANGLEY AFB	0600-2300 MON-FRI	125.0 253.5	121.7 275.8	271.8	PAR

Fig. 11-9. *The back of the sectional legend flap tells more about the tower, its frequencies, and hours of operation.*

```
ELIZABETH CITY CG AIR STATION/MUNI  (ECG)  3 SE   UTC-5(-4DT)                              WASHINGTON
    N36°15.63' W76°10.48'                                                              H-4J, 6H, L-22H, 27D
  12   B   S2   FUEL  80, 100LL, JET A1+   TPA—1012(1000)                                       IAP
 RWY 10-28: H7219X150 (CONC)    S-100, D-200, DT-400   HIRL   CL
   RWY 10: TDZL. REIL. VASI(V4L)—GA 3.0° TCH 58'. Road. Rgt tfc.
   RWY 28: REIL VASI(V4L)—GA 3.0° TCH 68'. Road.
 RWY 01-19: H4518X150 (ASPH-CONC)    S-20    MIRL
   RWY 01: VASI(V4L)—GA 2.75° TCH 31'. Thld dsplcd 299'. Road. Rgt tfc.
   RWY 19: VASI(V4L)—GA 2.75° TCH 28'.
 AIRPORT REMARKS: Attended 1200–0300Z‡. For attendant after 0300Z‡ and holidays call 919–335–5634 during
    hours attended. Intermittent tethered balloon ops 3 NM SE; dalgt hrs at 3000 ft. PAEW invof AER 19.
    Rwy 01-19 CLOSED to fixed wing tfc over 12500 lbs; vertical tkfs and ldgs only for helicopters over 12500 lbs.
    All DOD turbojet acft must obtain prior permission from air station ops prior to requesting arpt familiarization
    apchs. All apchs must be full stop lndgs. Coast Guard PPR for fuel and parking; call 919–335–6323. Coast
    Guard complex has numerous buildings within the building restriction lines. Strict compliance for noise
    abatement procedures are required in so far as possible, avoid overflying populated area NW below 1500 ft. acft
    departing rwys 01 and 28 expect climb to 1500 ft prior to turning. Blimp opr and training site 3 NM southeast of
    fld–Flight opr conducted in surrounding area. Seasonal low–flying agricultural acft based in the local area. No
    line-of-sight between rwy ends. General aviation ramp strength is considerably weaker than rwys. Twy B clsd
    indef. Twy C edge lgts are 50 ft from twy edge. Rwy 10 and 28 REIL OTS indef. When twr is clsd centerline lgts
    and VASI Rwy 10 and Rwy 28 opr continuously; REIL Rwy 10 and Rwy 28 are unavailable; ACTIVATE HIRL Rwy
    10–28 and MIRL Rwy 01–19 —CTAF.
 COMMUNICATIONS: CTAF 120.5
    RALEIGH FSS (RDU) TF 1–800–WX–BRIEF. NOTAM FILE ECG.
    RCO 122.2 122.05R 112.5T (RALEIGH FSS)
    WASHINGTON CENTER APP/DEP CON 123.85
    TOWER 120.5 (1200–0300Z‡)    GND CON 121.9
 AIRSPACE: CLASS D svc 1200–0300Z‡ other times CLASS G.
 RADIO AIDS TO NAVIGATION: NOTAM FILE ECG.
   (L) VOR/DME 112.5    ECG    Chan 72    N36°15.46' W76°10.54'    at fld. 10/07W.   HIWAS.
   WOODVILLE NDB (MHW) 254    LLW    N36°15.78' W76°17.87'    101° 6 NM to fld.
```

Fig. 11-10. *The A/FD provides even more communications information.*

The *A/FD* is published six times a year, each issue having approximately a two-month validity period. One advantage of this source is that, for a given location, it gives all the frequencies under the "Communications" heading: Unicom, Flight Service (either on the field or remoted), Approach/Departure Control, ATIS, Ground Control, Clearance Delivery, and tower (Fig. 11-10). The sectional is never this complete.

Just be sure you always use the current editions of the sectional and A/FD. Frequencies do change.

DOING WHAT THE TOWER TELLS YOU

Yes, you are the pilot in command, with certain responsibilities and authority. And, yes, the tower is there to serve you, along with all the other pilots in the area. Neither condition, however, alters the fact that no pilot has the right to go his way regardless of the tower's instructions. To repeat what I've already said, unless an emergency suddenly develops or adherence to an instruction would violate a FAR, you must comply with the tower's directives.

The controller is spacing, separating, coordinating, and overseeing all the traffic within his or her area of responsibility. He or she has a plan to get everyone up or down with minimum delays and maximum safety. The whole operational pattern can't be disrupted because someone does a 360 on the downwind leg or chooses to land on Runway 21 when Runway 18 is the active runway. Perhaps judgment decrees that a 360 is essential for spacing or safety. If so, advise the tower before starting the maneuver. Perhaps Runway 21 is better because of the wind. Fine, but get permission before you switch to another pattern.

Does any of this sound fundamental? Sure, but such unannounced or unapproved deviations from instructions are not rare. Stick around a controlled airport for a while. Listen and observe. You'll see.

THE TAKEOFF CONTACT

The start of this chapter had you ready for takeoff. Let's continue from there. Assume that you're number one to go and are still in the runup area, some 75 feet or so from the hold line. After the pretakeoff checks are complete, taxi to the hold line and come to a complete stop. At this point, you're clear to switch from Ground Control to the tower frequency. Before making the call, however, check your volume, and be sure you're ready to transmit as well as receive on the proper frequency.

Why the call at the hold line (Fig. 11-11) and not back in the runup area? Simple. Suppose an aircraft on a relatively short base is about to turn for final approach. If you're back there in the runup area, the tower would almost certainly tell you to "taxi closer and hold." There just wouldn't be time to taxi from where you are to the runway and get rolling before the landing craft would be on your tail. On the other hand, if you were at the hold line, the tower might be able to clear you for immediate takeoff. Through this simple maneuver, traffic is expedited and aircraft on the ground behind you have less idling- and fuel-burning time.

When you're really ready to go, the call is nothing more than this:

You: Cedar Tower, Cherokee One Four Six One Tango ready for takeoff, east departure.

Tower: *Cherokee Six One Tango cleared for takeoff. East departure approved.*

You: Roger, cleared for takeoff, Cherokee Six One Tango.

Remember KISS. No need for something like this:

You: Cedar Control Tower, this is Cherokee One Four Six One Tango, Over.

Tower: *Cherokee One Four Six One Tango, Cedar Tower.*

You: Cedar Control Tower, this is Cherokee One Four Six One Tango. We're ready to take off on Runway Three Six. We would like to leave the pattern and depart to the east. Over.

Tower: *Cherokee Six One Tango cleared for takeoff, east departure approved.*

You: Roger. Six One Tango. East departure approved.

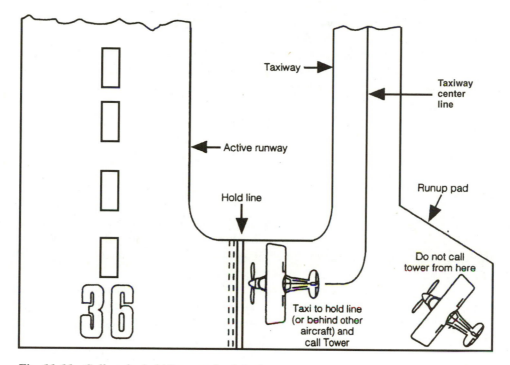

Fig. 11-11. *Call at the hold line, not back in the runup area.*

That transmission consumed 52 words, not counting the tower's replies. What have you said that you didn't say in 21 words in the other departure call? Nothing. And the tower knows you're going to use Runway 36, so why state the obvious?

Now let's assume that you've made the call and the tower acknowledges with: "Cherokee Six One Tango, hold for landing traffic." Your reply should indicate your understanding and compliance: "Six One Tango holding."

The landing aircraft has landed, and the tower contacts you again: "Cherokee Six One Tango, taxi into position and hold." Your reply: "Position and hold, Six One Tango."

You haven't merely "Rogered" the instruction. You've confirmed to the tower that you have understood and will comply with the instructions.

When the landing craft has cleared the runway, the tower calls you once more: "Cherokee Six One Tango, cleared for takeoff." What's missing? Should you take off under the possibly false assumption that your eastbound departure has been approved? No, and don't assume anything. Contact the tower again:

You: Tower, Cherokee Six One Tango. Is eastbound departure approved?

Tower: *Cherokee Six One Tango, Roger, eastbound departure approved. Remain north of the buildings.*

You: Wilco ["Will comply"], Cherokee Six One Tango.

Another situation: After your initial "ready for takeoff" call, the tower says: "Cherokee Six One Tango cleared for immediate takeoff."

What's the controller saying? He's telling you to get out on that runway and go. "Immediate" means *immediate*. You don't idle your way to 36 and take your sweet time setting everything up. Taxi out as rapidly and safely as you can to the center of the runway, line up the airplane, and apply power *now*.

This sort of instruction doesn't require your acknowledgment. The fact that your aircraft is moving is indication enough that you have received the message. If you sit on the runway in takeoff position with no visible activity on your part, you're certain to hear from the person in the elevated enclosure.

A variation of this situation: Responding to your ready-to-go call, the tower comes back with: "Cherokee Six One Tango cleared for immediate takeoff or hold short." It's decision time. If you're really ready, taxi out and get going; the rolling aircraft confirms your intentions. Otherwise, tell the tower what you're going to do. Silence leaves the controller in a state of uncertainty—which makes controllers very unhappy. Don't play *I've Got a Secret*. Quickly reply, "Cherokee Six One Tango will hold."

One other example of these takeoff contacts: Ahead of you at the hold line are two other aircraft. You've completed your runup and are ready to go, but the other planes are just sitting there. If, for example, they're waiting for IFR clearances, they might remain sitting for some time, while you stay patiently in line burning up fuel. So what do you do? The logical thing is to call the tower. If possible, the tower will clear you ahead of those who precede you:

You: Cedar Tower, Cherokee One Four Six One Tango, number three in sequence, ready for takeoff. Request east departure.

Tower: *Cherokee Six One Tango. Taxi around the Cessna and Mooney. Cleared for takeoff. East departure approved.*

You: Roger, cleared for takeoff, Cherokee Six One Tango.

CHANGING FREQUENCIES AFTER TAKEOFF

For a variety of reasons, you might want (or need) to change frequencies shortly after taking off. Perhaps you're going to a nearby airport and want to monitor the traffic; your departure route takes you over another airport and clearance to cross it is required; you want to contact Departure Control; you need to open or modify your VFR flight plan with Flight Service; and so on. Whatever the reason, always ask for and receive approval from the tower to change to another frequency as long as you are still within the Class D area—the 5 mile radius. The request is: "Cedar Tower, Cherokee Six One Tango requests frequency change."

In all likelihood, the tower will approve the request. If local traffic is heavy or visibility limited, however, the controller might want you to stay with him until you are well clear of his area. Depending on the circumstances, his response might be: "Cherokee Six One Tango, frequency change approved," or "Cherokee Six One Tango, remain this frequency. I'll have traffic for you," or "Cherokee Six One Tango, stay with me until clear of the area."

The point is: Don't leave the tower while within the Class D airspace until the tower has approved the change. The controller might need to contact you for any number of reasons, and you are still within airspace under his control.

TAKEOFF WITH CLOSED PATTERN

This time, you want to sharpen your landings with a few touch-and-gos:

You: Cedar Tower, Cherokee One Four Six One Tango, ready for takeoff. Request closed pattern.

Tower: *Cherokee One Four Six One Tango Roger. Cleared for takeoff. Closed pattern approved.*

From this point on, you usually won't have to initiate any further contacts with the tower. This is a controlled airport, and the tower knows your intentions. All that's necessary is to acknowledge the controller's transmissions to you:

Tower: *Cherokee Six One Tango, cleared for touch-and-go, Runway Three Six.*

You: Roger. Cherokee Six One Tango.

Again, however, keep the tower informed. When you decide that you've had enough practice, it's a good idea on the downwind leg to tell the tower that this will be a full stop—even before the controller clears you for another touch-and-go. Initiating the call early on the downwind leg might help the controller space other aircraft that are either landing or taking off:

You: Cedar Tower, Cherokee Six One Tango, will be full stop this time.

Tower: *Cherokee Six One Tango, Roger. Cleared to land, Runway Three Six.*

Whether you initiate the call or respond to the controller's clearance for another touch-and-go, be sure to inform him of your intentions. Controllers don't like surprises. However, unlike at multicom or unicom airports, do not call the tower on downwind or when turning base and final, unless there is a special reason to do so. The controller knows what you're doing, and these calls only clog the air.

APPROACHING-THE-AIRPORT CONTACTS

You're on a cross-country flight and are nearing your destination airport. Just to make it simple, let's say that there's no Class B or C associated with the field. Be a good Scout before making the initial contact: Be prepared!

An airline captain told this story of what came over the air one time. As he was approaching a field, he heard a charter carrier talking to the tower. With names and locations changed to protect the guilty, the call went something like this:

Pilot: Mayflower Tower, this is Rocky Charter Three Niner Niner Uniform.

Tower: Rocky Charter Three Niner Niner Uniform, go ahead.

Pilot: Tower, Three Niner Niner Uniform, we're over…[to the first officer, with an open mike] Where are we, Harry? [Pause] We're over North Centerville for landing at Mayflower.

Tower: Roger, Three Niner Niner Uniform. Say altitude.

Pilot: Altitude is…What's our altitude, Harry? [Pause] Altitude is seven thousand five hundred.

Tower: Roger, Three Niner Niner Uniform. What's your airspeed?

Pilot: Let's see, airspeed…how fast we going, Harry? (Pause) 375 knots, Tower.

Tower: What are you squawking, Three Niner Niner Uniform?

Pilot: We're squawking…Harry, what are we squawking?

At this point, the tower broke in:

Tower: Three Niner Niner Uniform, would it be all right if we talked to Harry?

Not having personally heard this exchange, I can't vouch for its authenticity. The airline captain, however, left no doubt in my mind that it was almost a verbatim copy of what transpired.

Know what you're going to say, and if you're a little uncertain, rehearse it to yourself before you pick up the mike and press the button. Also, if you've set things up properly, you've listened to the ATIS and monitored instructions to other aircraft on the tower frequency. With these transmissions, you know the winds, altimeter setting, active runway, and traffic pattern direction. Now you're equipped to plan your approach because you have a good idea what the tower will tell you when you establish contact.

At what point should you make the initial call to the tower? To repeat: Radio contact with ATC must be established before you enter the Class D airspace, so a reasonable rule of thumb is to introduce yourself about 15 miles out. If you've monitored the frequency as just suggested, you should know how much traffic the tower is handling and how busy it is. To illustrate a couple of situations, though, let's assume for the moment that the activity is high. Your call, then, should be designed only to establish that first contact. Example:

Pilot: Cedar tower, Cherokee One Four Six One Tango.

Tower: Cherokee One Four Six One Tango, Cedar tower: Go ahead.

Fine. Contact has been established; you have the controller's attention and can now provide the rest of the information the tower needs. We'll come to that shortly, but let's suppose the tower replies to your call with the following:

Tower: Cherokee One Four Six One Tango, stand by.

With that response, and even if you don't hear from the tower again for several minutes, have you been cleared to enter the Class D airspace? The answer: "Yes." The tower has acknowledged your presence by addressing you by your call sign.

On the other hand, what if the tower came back with this response?

> **Tower:** *Aircraft calling Cedar tower, remain clear of the Delta airspace and stand by.*

or

> **Tower:** *Aircraft calling Cedar tower, stand by.*

In either case, the answer is obvious. You have not been cleared to enter the airspace, so avoid all portions of it until the tower gets back to you. Do whatever you have to do— circle, wander around, do loops—it's your choice. But you'd better keep your eyes open for other traffic. If ATC can't clear you into the airspace, the odds are good that there's considerable activity throughout the neighboring vicinity. Meanwhile, to be sure that the tower knows that you have understood its directive, briefly acknowledge the call:

> **Pilot:** Roger, Cedar. Cherokee Six One Tango remaining clear.

Going back to the information about you that the tower needs, whether you submit it as part of the very first call or in a followup contact, this is what you should communicate, and, to make it simple for you, communicate it in the sequence suggested.

1. *I*dentification (meaning aircraft name, type, and N-number)
2. *P*osition (where you are geographically in relation to the airport)
3. *A*ltitude
4. *I*ntentions/*D*estination (example: "landing Cedar Airport," or "touch-and-gos Cedar)
5. *S*quawk code: [as "squawking twelve" (for 1-2-0-0)]
6. *A*TIS information

It's not a perfect acronym, but I-P-A-I/D-S-A can be a memory-jogger that, if followed, will give the tower all the information it needs to accept you into the airspace and to coordinate your sequence into the traffic pattern.

Assuming that the tower does not appear to be overly busy, the first call, then, would go like this, without necessarily taking up the time to establish that initial contact suggested above:

> **Pilot:** Cedar tower, Cherokee One Four Six One Tango over Grand Lake, level at three thousand five hundred, landing Cedar. Squawking twelve, with Lima.

> **Tower:** *Roger, Cherokee One Four Six One Tango, enter left downwind for landing Runway one eight.*

> **Pilot:** Roger, Cedar. Left downwind for One Eight. Six One Tango.

Chapter Eleven

Once again, I've got to stress the importance of tersely repeating ATC's instructions. The controller wants to know that you know what he or she expects you to do. You're told to "Enter right base for two two." Acknowledge with "Roger. Right for two two." Or your instruction is to "Cross midfield at two thousand three hundred for left downwind." The response: "Roger, midfield at two thousand three hundred, left downwind." Or you're asked to report 4 miles east at the twin bridges. "Roger, four east at bridges. Six One Tango."

Are the pilot and controller on the same page in the same hymn book? They'd better be. Brief summaries or instructions should ensure that the two parties are in harmony.

Probably not many of us want to clutter up the air with needless chatter or time-consuming commentaries, but, as was the case of a pilot in one of the accidents cited in Chap. 2, suppose that part of the tower's transmission was garbled or indistinct and you think the active runway is 24 when it's actually 34. If you proceeded to plan your entry into the airspace and the pattern accordingly, it might have some serious disruptive effects. Obviously, such confusion shouldn't exist at all going into any Class D airspace— if you've monitored the ATIS and kept listening to the tower and its contacts with other aircraft in the area. But, these things do happen.

As just one example, I heard a Cessna 172 some time ago that was piloted by Lieutenant Confusion or Captain Stupid. It was coming into the Kansas City Downtown Airport from the south, with 01 as the active runway. After I heard the Cessna first on Approach Control, I switched to the tower to announce my position. A few minutes later, the Cessna turned up well north of the airport and hesitantly contacted the tower. The tower clearly identified the active as Runway 01, with right-hand traffic. The Cessna dutifully rogered the information. Next, the tower told the Cessna to fly a heading of 190 degrees on the downwind. Again, the Cessna rogered. Apparently, the tower was keeping a close eye on the errant aircraft, because the next question was: "Cessna Zero Zero Zero Zero, are you landing on Runway Three?" To which the Cessna rogered again.

Now, Runways 01 and 03 happen to cross each other, and cross-traffic landings and takeoffs, if not controlled, can lead to messy situations. Fortunately for at least two aircraft, traffic was sufficiently spaced and the controller sufficiently alert to permit the visiting Cessna, which was now on a tight base, to continue to Runway 03. No harm was done.

Whether Confusion or Stupid was in command doesn't matter. It is unlikely that Approach vectored the pilot over and well north of the airport when a straight-in-approach from the south to Runway 01 was the shortest distance between the two points. The pilot must have confused Runway 03 with Runway 01, and he was flying a downwind leg at 210 degrees instead of the required 190 degrees. And he calmly rogered the question, "Are you landing on Runway Three?"

If the Cessna pilot had rogered less and repeated at least a couple of the instructions just once, the tower could have straightened things out before a potential hazard had arisen. "Roger" doesn't necessarily ensure commonality of understanding.

Let's continue with the approach and landing. If you're transponder-equipped, you might receive no further instructions from the tower other than"Cherokee Six One Tango, cleared to land." To which you reply "Roger, cleared to land, Cherokee Six One Tango." Roger. You've got it.

On the other hand, and particularly if you have a transponder, the tower might be in frequent communication to advise you of other aircraft in your vicinity:

Tower: *Cherokee Six One Tango, traffic is a Cessna at one o'clock, two miles, westbound at two thousand.*

[Pause while looking]

You: Negative contact, Cherokee Six One Tango.

Or, if you spot the traffic: "Cherokee Six One Tango has the Cessna."

Another situation: you're advised of the Cessna at the one o'clock position. You don't see it, and so inform the tower. A minute or so later, you spot it at your two o'clock position. At this point, tell the tower—even though the traffic is well to your right and presents no possible hazard: "Cedar Tower, Cherokee Six One Tango has the Cessna." The tower will acknowledge your message, often with a "thank you."

In a similar scenario, you hear the tower call another aircraft in your general area, alerting the pilot to your presence: "Cessna Eight Niner Golf, traffic is a Cherokee at ten o'clock, two miles, also westbound. Altitude unknown."

You have a reasonably good idea that that's you, so get on the air and help everybody by reporting your altitude: "Cedar Tower, Cherokee Six One Tango is at two thousand eight hundred, descending to one thousand seven hundred (pattern altitude)." You might not get an acknowledgment, but that's beside the point. You've kept the other parties informed—at least the interested parties.

This is as good a time as any to mention that when you're given traffic in a "clock" position (nine o'clock, one o'clock, and so forth), the position is based on your *ground track,* as shown on radar. The traffic's position might be different from your point of view, because of your wind correction angle. For example, you're tracking over the ground at 270 degrees, but because of a northerly wind, your heading is 300 degrees. If you are advised that you have a "target" at "12 o'clock," that means that the other aircraft is on your 270-degree track, not straight ahead of the nose of your airplane at 300 degrees. The target, then, is actually at about 11 o'clock in relation to the direction in which your airplane is pointed. Keep this in mind, especially under strong wind conditions when you need to crab to maintain your desired course.

Now let's continue and say you're nearing the airport. Once again, the tower calls you:

Tower: *Cherokee Six One Tango, you're number two to land behind the Mooney on downwind.*

You: Roger, number two to land. Negative contact on the Mooney [or "and we have the Mooney"]. Cherokee Six One Tango.

When the spacing is proper between you and the Mooney, or the Mooney is about to touch down, the tower will give you final clearance:

> *Tower:* *Cherokee Six One Tango, cleared to land, Runway Three Six.*
>
> **You:** Roger, cleared to land Three Six, Cherokee Six One Tango.

Once you're on the ground, be sure to stay on the tower frequency until you have turned off the active runway and the tower has cleared you to the ramp or advised you to contact Ground Control. You have no way of knowing what might be happening behind you that would require the tower to issue you an emergency instruction. The call from the tower will probably be brief—no more than "Cherokee Six One Tango, contact Ground point niner." Remember that the Ground Control frequency is typically 121.7, 121.8, or 121.9. Consequently, the first three digits are often dropped.

Occasionally, especially if the tower is busy, the controller might fail to tell you to contact Ground, even though you are clear of the runway. In such cases, go past the taxiway hold line, come to a complete stop, and call the tower:

> **You:** Tower, Cherokee Six One Tango going to Ground.
>
> *Tower:* *Cherokee Six One Tango, Roger. Contact Ground.*

This clears you to leave the tower frequency. Remember, stay on the tower frequency until a change is authorized.

OTHER TRAFFIC PATTERN COMMUNICATIONS

Because of spacing, the tower wants greater separation between you and the aircraft ahead of you:

> *Tower:* *Cherokee Six One Tango, extend your downwind for spacing.*
>
> **You:** Roger. Cherokee Six One Tango. Will you call my base? [Meaning, "Will you tell me when I can turn to the base leg?"]
>
> *Tower:* *Cherokee Six One Tango, affirmative.*
>
> *Tower:* *Cherokee Six One Tango, turn to base. You're number two to land behind the Aero Commander on final.*
>
> **You:** Roger, and we have the Commander [or "negative contact on the Commander"], Cherokee Six One Tango.

You're on final, 100 feet above touchdown, and an unauthorized aircraft or ground vehicle ventures onto the runway. To avoid a confrontation, the tower issues a command:

> *Tower:* *Cherokee Six One Tango, go around!*
>
> **You:** [No response is necessary. Pour on the coals and initiate the go-around procedure. Don't argue; don't debate. Just do what you're told.]

You've been shooting touch-and-gos but would now like the option of making a touch-and-go, stop-and-go, or a full-stop landing. Make the request on the downwind leg so the tower can say yea or nay, based on the existing traffic:

You: Cedar Tower, Cherokee Six One Tango requests the option.

Tower: *Cherokee Six One Tango, cleared for the option.*

The option approach is especially useful during flight instruction to maintain an element of surprise for the student, because go-arounds and missed instrument approaches are also permitted as options.

On final approach, decide (or have your instructor decide) how you'll end the approach. No need to tell the tower. You're cleared for whatever you decide.

When you've had enough for the day, advise the tower of your intentions—again on the downwind leg:

You: Cedar Tower, Cherokee Six One Tango will be full stop this time.

Tower: *Cherokee Six One Tango cleared to land.*

You: Cleared to land, Cherokee Six One Tango.

At this point, the tower makes a request of you:

Tower: *Cherokee Six One Tango, can you land short and turn off on Runway Two One?*

You: Affirmative, Cherokee Six One Tango. [Assuming you can comply with the request.]

Don't just come back with a "Roger." Can you or can't you? The response is either "Affirmative" or "Negative."

A TOWER BUT NO GROUND CONTROL OR ATIS ON THE FIELD

As I indicated in Chap. 9, most tower-controlled airports provide Ground Control service. At many of the smaller and less active fields, however, the tower controller is often responsible for both ground and flight movements. Since the sectional charts themselves do not indicate the presence or absence of the service, you have to refer to the table on the inside of the sectional cover flap. If you're at a tower-operated field that does not have Ground Control, merely contact the tower. The controller will then provide the necessary taxi instructions.

Occasionally, even when there is Ground Control, this can happen: You've just landed and are clear of the runway. Rather than telling you to contact Ground Control, the tower assumes the taxiing-control responsibility.

Tower: *Cherokee Six One Tango, taxi to the ramp.*

Or if there are several locations where you could park, it might be this:

Tower: *Cherokee Six One Tango, where do you want to park?*

You: Hangar Four, Tower.

> *Tower:* *Roger, Six One Tango, taxi to Hangar Four.*
>
> **You:** Roger, Six One Tango.

The assumption of the Ground Control role by the tower—if it does happen at all—occurs when the area traffic is light and the controller has the time to issue brief instructions. Just don't be surprised if the tower assumes the responsibility. The possibility of it, however, is another reason why you don't change frequencies without ATC approval.

As to the ATIS not being on the field: You know from Chap. 8 that both its presence and frequency are indicated on the sectional and in the *A/FD*. In general, that tape-recorded service is available at most, but not all, tower-controlled airports.

Where you are more likely not to find ATIS is at those towers manned by nonfederal contractors. These are "civilians," if you will, who represent private firms that have contracted with city governments to provide the traffic-controlling service at the local airports. In the vast majority of such cases, and for whatever reason, ATIS is not part of that service. Frequently, however, these "NFCT" (Non-Federal Control Tower) fields are near major airports that do have ATIS that pilots can monitor for a fairly accurate picture of conditions at whatever non-ATIS field they're going to operate in.

Fine, but how do you know if it's a nonfederal tower? Again, the sectional chart tells, both on the chart itself by the airport symbol and on the inside cover flap. On the chart, the identification is "NFCT," just preceding the tower frequency. If you'll check the Manhattan, Kansas, example in Fig. 11-12, you'll see what I mean. On the legend flap (Fig. 11-13), "NF" follows the airport name.

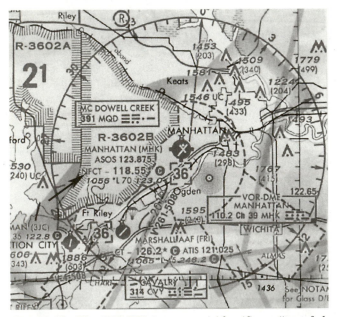

Fig. 11-12. The "NFCT" (see arrow) identifies a "non-federal control tower."

CONTROL TOWER FREQUENCIES ON KANSAS CITY SECTIONAL CHART

Airports which have control towers are indicated on this chart by the letters CT followed by the primary VHF local control frequency. Selected transmitting frequencies for each control tower are tabulated in the adjoining spaces, the low or medium transmitting frequency is listed first followed by a VHF local control frequency, and the primary VHF and UHF military frequencies, when these frequencies are available. An asterisk (*) follows the part-time tower frequency remoted to the collocated full-time FSS for use as Local Airport Advisory (LAA) during hours tower is closed. Hours shown are local time. Ground control frequencies listed are the primary ground control frequencies.

Automatic Terminal Information Service (ATIS) frequencies, shown on the face of the chart are primary arrival VHF/UHF frequencies. All ATIS frequencies are listed below. ATIS operational hours may differ from control tower operational hours.

ASR and/or PAR indicates Radar Instrument Approach available.

"MON-FRI" indicates Monday thru Friday.

CONTROL TOWER	OPERATES	TWR FREQ	GND CON	ATIS	ASR/PAR
BILLARD NF	0700-1900	118.7 257.8	121.9		
COLUMBIA REGIONAL	0700-2300	119.3*	121.6	128.45	
DRAKE	0600-2200	128.0 371.9	121.8	133.1	ASR
FORBES	0545-2200	120.8 255.9	121.7 275.8	128.25	
FORNEY AAF	0830-1630 MON-FRI EXC HOL	125.4 241.0		118.7 237.5	
JEFFERSON CITY MEM NF	0600-2130	125.6	121.7		
JOHNSON CO EXECUTIVE	0700-2100	126.0 241.1	121.6	119.35	
JOPLIN REGIONAL	0600-2100	119.8 320.1	121.6	120.85	
KANSAS CITY DOWNTOWN	CONTINUOUS	133.3 257.8	121.9	120.75	
KANSAS CITY INTL	CONTINUOUS	128.2 254.25	121.8	128.35	
LAMBERT ST LOUIS INTL	118.5/257.7 (SOUTH) 120.05/284.6 (NORTH) 0600-2300 120.05/289.1 2300-0600	118.5 120.05 257.7 284.6 289.1	121.9 348.6	120.45 277.2	
MANHATTAN NF	0800-1800	118.55	121.85		
MARSHALL AAF	0800-2400 MON-FRI CLSD HOL	126.2 248.2	121.7 229.4	121.025	ASR/PAR
NEW CENTURY NF	0600-2200	118.3	124.3		
QUINCY-BALDWIN	0800-1600 SAT & SUN FIRST WKND EV	135.25 234.875	121.75 238.325		

Fig. 11-13. The back of the sectional chart's legend flap also identifies a non-federal control tower.

If you're at an NFCT field and there's no ATIS, the tower will give you the basic operational data—wind, direction, velocity, altimeter setting, and the active runway. That's all, barring unusual weather conditions in the airport vicinity. You now have "the numbers" essential for a takeoff or a landing. To get this data, simply request it in your initial call:

You: Billard Tower, Cherokee One Four Six One Tango is over Perry Dam at four thousand five hundred for landing. Request the numbers.

Tower: *Cherokee One Four Six One Tango, enter left base for Runway One Eight. Winds two zero zero at one five. Billard altimeter three zero one two. Report entering base leg.*

You: Roger, report entering left base for One Eight. Cherokee Six One Tango.

The procedure is exactly the same prior to taxiing for takeoff when there is no ATIS. Merely tell Ground (or the tower, if there is no Ground Control) where you are, that you're ready to taxi, followed by "Request the numbers." Either will give you the essential data.

OBTAINING SPECIAL VFR CLEARANCE

Back in Chap. 7, I discussed obtaining Special VFR clearances when operating within a Class E airspace under less than VFR conditions. Now let's assume that the same conditions exist at a Class D airport with its operating tower. Barring only minor variations, the procedures and radio calls are basically the same as those when a Flight Service Station is involved.

First, listen to the ATIS and then call Ground Control and request the SVFR:

You: Municipal Ground, Cherokee One Four Six One Tango at Jet Air with Information Kilo. Request Special VFR southbound.

GC: *Cherokee One Four Six One Tango, taxi to Runway Three Three. Clearance on request.* [This means that Ground Control is requesting the clearance for you from ATC—Approach/Departure Control or Center. It does not mean that the clearance is available to you on your request.]

You: Roger. Taxi to Three Three. Cherokee One Four Six One Tango.

GC: [In a few minutes] *Cherokee One Four Six One Tango, clearance when ready to copy.*

You: [Assuming you're free to copy] Cherokee Six One Tango ready.

GC: *Six One Tango, ATC clears Cherokee One Four Six One Tango out of the Delta airspace to the south. Maintain Special VFR conditions at or below two thousand while in the Delta airspace. Report leaving the airspace.*

You: Understand Cherokee One Four Six One Tango cleared Special VFR out of the Delta airspace to the south, maintain two thousand or below while in the airspace, and report when clear of Delta.

GC: *Readback correct, Six One Tango. Contact tower.*

You: Roger, Cherokee Six One Tango.

When you've finished the pretakeoff check and are ready to go, switch to the tower, which will already be aware of your clearance. Thus the following:

You: Municipal Tower, Cherokee One Four Six One Tango ready for takeoff, Special VFR clearance southbound.

Tower: *Cherokee One Four Six One Tango, cleared for takeoff. Report when clear of the Delta airspace.*

You: *Will do, Six One Tango.*

Now let's say you want to land at Municipal Airport instead of departing. Through ATIS, you find that the field is below VFR limits but adequate for a Special VFR. Consequently, when you're several miles out from the Delta airspace, you call the tower:

You: Municipal Tower, Cherokee One Four Six One Tango is over Deep Lake at three thousand, squawking one two zero zero with Information India. Request Special VFR for landing Municipal.

Tower: *Cherokee One Four Six One Tango, roger. Remain clear of the Delta airspace until further advised.*

Now circle, slow down, or do whatever is necessary to stay outside the airspace until you hear from the tower again.

Tower: *Cherokee One Four Six One Tango, Municipal Tower. Clearance when ready to copy.*

You: Roger, Six One Tango ready.

Tower: *Six One Tango, ATC clears Cherokee One Four Six One Tango to enter the Delta airspace east of Municipal. Maintain Special VFR conditions at or below two thousand five hundred. Squawk one two five zero. Report right base for Three Five.*

You: Roger, Tower. Understand Cherokee One Four Six One Tango cleared to enter the Delta airspace east of Municipal. Maintain Special VFR at or below two thousand five hundred. One two five zero, report right base for Three Five.

Tower: *Readback correct, Six One Tango.*

And sometime later:

You: Tower, Cherokee One Four Six One Tango turning right base for Three Five.

Tower: *Roger, Cherokee One Four Six One Tango, cleared to land. Winds three two zero at five.*

That's all it takes to request, and hopefully obtain, Special VFR clearances out of or into Delta airspaces. Of course, if the airport is in a Bravo or Charlie airspace, you'd contact Approach Control for clearance to enter the airspace. Then, at the proper time, Approach would tell you when to call the tower for landing clearance.

A FEW WORDS ABOUT DEVIATIONS

Something I've noticed, particularly with newer pilots, is the reluctance to ask the tower for traffic pattern deviations when another runway or approach is more practical. There is no emergency, no danger of violating a VFR regulation, but a different pattern would be more practical or time-conserving. While the reluctance is understandable, controllers occasionally authorize such deviations if the volume of traffic permits. To be more specific:

There are two runways on the field, one is 19/01, the other 21/03. You've already learned from monitoring the ATIS that One Nine is the active runway, but the wind is from 220 degrees at 15 knots, gusting to 25. That could present a fairly stiff crosswind, so you decide to ask the controller in your initial contact if 21 can be used:

You: Downtown Tower, Cherokee One Four Six One Tango is over Lake Quivera at three thousand for landing, squawking one two zero zero with Information Charlie. Is Two One available?

Tower: *Affirmative, Cherokee One Four Six One Tango. Ident and enter left downwind for Two One.*

You: Roger, left for Two One. Cherokee Six One Tango.

Did you push the Ident button, as the controller asked?

Another possibility:

You're coming in from the west and, again from the ATIS, you learn that One Nine is the active runway with a left pattern. To enter the downwind for that pattern, though, means you'd have to fly over the field, go east a few miles, descend, and then join the other downwind traffic at a 45-degree angle. All of which is both time- and fuel-consuming. ATC might approve a direct entry to a right base leg almost dead ahead of your present position. Thus the query:

You: Downtown Tower, Cherokee One Four Six One Tango is over Wyandotte Lake at two thousand five hundred for landing, squawking one-two-zero-zero with Information Delta. Request right base, if possible.

Tower: *Roger, Cherokee One Four Six One Tango, Ident, and report entering right base for One Niner.*

You: Roger, report right base for One Niner. Cherokee Six One Tango.

A few minutes later:

You: Tower, Cherokee Six One Tango on right base for One Niner.

Tower: *Cherokee Six One Tango, cleared to land.*

You: Six One Tango cleared to land.

If the tower can't grant your request, no problem. The controller will then come back with something like this:

Tower: *Unable right base, Cherokee Six One Tango. Ident and fly over the field at two thousand three hundred for a left downwind for One Niner.*

Fine, but at least you tried. Now merely acknowledge the instructions and do what the tower has said.

Regardless of the airspace, controllers can be flexible when conditions permit. You know where you are and what's best for you, so don't be afraid to request a reasonable deviation. Just be sure, though, that it's reasonable. No one in his or her right mind would ask to land on Runway 01 at a busy airport when the wind is out of the south and all the traffic is using 19. Let's hope not, anyway.

CONCLUSION

Remember, the communications examples cited in this book basically conform to the FAA-approved phraseology. It's appropriate to note, however, that you'll occasionally hear some verbal shorthand that might be acceptable but not necessarily correct by FAA standards.

For example, when approaching an airport for landing, a pilot may be heard "Squawking one two zero zero" less and less—if at all in the initial call. Or it might be reduced to "Squawking Twelve." Pilots—not controllers—sometimes shorten altitudes from "Level at two thousand three hundred" to "Level at two point three." The fact that the pilot has "…Information Alpha" is often shortened to "…with Alpha." The word "Information," being somewhat self-evident, is dropped entirely. Similarly, B, C, or D airspaces are referred to simply as "Bravo," "Charlie," or "Delta," versus "Class B airspace," and so on. And there are others. I recommend, though, that you master the approved radio language first. Get that down before the tendency to slip into some of the typical but incorrect pilot jargon becomes too inviting.

To sum up, and as I've tried to illustrate in this chapter, there's nothing complicated or mysterious about operating in a Class D airspace. From a radio communications point of view, it's simply a matter of adhering to a few basic principles, the majority of which can be condensed this way:

Monitor the ATIS 20 miles or so out from the airport.

Monitor the tower frequency ahead of time to learn what you can about the traffic pattern and the volume of activity in the area.

Plan what you're going to say and how you're going to say it.

Listen before you press the mike button to be sure the air is clear.

Listen to what the controller tells you.

Clarify instructions, if you're not sure.

Listen for your call sign even after you've been given landing pattern instructions. The tower might have additional instructions for you.

Acknowledge instructions briefly and tersely.

Inform the controller of what you're doing or going to do if a deviation is required.

Request deviations from instructions if there is a more effective or efficient nonemergency alternative.

Communicate with confidence.

I know that I'm repeating myself, but you're a team out there—pilots and controllers. When both function in unison, especially in the usually busy environment of controlled airports, traffic flows as it should. And when it flows properly, the folks up in those towers can be the nicest folks around. Just remember, though: They can do without you, but you can't do without them. So help them help you.

12

Operating
and Communicating
in Class B, C,
and TRSA Airspaces

BACK IN 1986, THE CERRITOS, CALIFORNIA, MIDAIR COLLISION between a small aircraft and an airliner near the Los Angeles Airport instigated, or accelerated, a flurry of FAA regulatory changes designed to prevent recurrences of such an incident. The changes, primarily affecting the major airports and surrounding airspaces, involved their structures, the adoption of stricter Mode C requirements, and more stringent operating regulations within those airspaces. Two of the products of these changes, along with the 1993 airspace reclassification rulings, are the Class B (Bravo) and Class C (Charlie) airspaces. These both primarily affect operations in and around the immediate airport environment.

The thrust of this chapter, then, is to review the structure of and the operating requirements in the Class B and C airspaces, along with a few comments about that neither-fish-nor-fowl Terminal Radar Service Area (TRSA). In the process of this review, I'll inevitably refer to the Approach and Departure Control function, as I did in Chap. 11. I ask, though, that these references not confuse you, because I'll discuss that function, its responsibilities, and the correct related radio procedures in Chap. 13. Let's just say for now that when you contact Approach or Departure, you're not talking to someone in that glass-enclosed facility, or "cab," atop the airport tower. Instead, the individual responding to your call will be located in a windowless room, probably a few floors below the cab itself. There, in semidarkness, he or she, and a team of fellow specialists, are responsible for the radar control of all traffic leaving, arriving, or transiting the immediate airspace *except* that traffic which is within the approximate five-mile radius of the airport itself and is thus under jurisdiction of the control tower.

If you're inexperienced or new to this business of flying, the thought of going into one of the Bravo or Charlie airspaces might be a bit intimidating. Should such be the case, the temptation to avoid these "big" airports in favor of those that are smaller or less convenient could be considerable. In fact, it might be overwhelming. Intimidation, though, is often the result of insecurity born of uncertainty or lack of knowledge. Natural though it is, insecurity can be overcome, first with knowledge and second with practice. These pages can't provide the practice you may need, but they hopefully can contribute to the acquisition of knowledge so that confidence replaces unjustified intimidation. If that should be the ultimate outcome, then it's unlikely that there's a Class B anywhere that would intimidate you or that you would avoid out of fear. (I do admit, though, that common sense on a VER flight might suggest bypassing one of those high-density, superactive airports such as O'Hare, Kennedy, or Atlanta's Hartsfield at five o'clock in the afternoon).

Having said that, let's focus now on the Class B airspace, which is the more controlled and, because of the nature of the traffic, certainly the busiest airspace of them all.

THE CLASS B AIRSPACES: WHAT THEY ARE AND WHERE THEY ARE

The traditional description of a Class B, as mentioned back in Chap. 4, is that of an upside-down wedding cake, composed of a central core with succeedingly higher layers or levels extending outward from the core.

Figure 12-1 is a generalized profile of a Class B structure, while Fig. 12-2 is a top view. These are simplistic sketches because Bravos are not always as neat and cylindrical, nor are they of the same design from airport to airport. Nevertheless, the figures do reflect the fundamental concept. For a real-life example, refer to Fig. 12-3, the Kansas City Class B airspace.

At the time of writing, there were 33 Class Bs located at the following largest and busiest airports. Those with an asterisk (*) identify the twelve busiest Bravo airports which also have special pilot requirements.

Operating and communicating in Class B, C, and TRSA airspaces

Atlanta*	Houston	Orlando
Baltimore*	Kennedy*	Philadelphia
Boston*	Kansas City	Phoenix
Charlotte	LaGuardia*	Pittsburgh
Chicago*	Las Vegas	St. Louis
Cleveland	Los Angeles*	Salt Lake City
Dallas/Ft. Worth*	Memphis	San Diego
Denver	Miami*	San Francisco*
Detroit	Minneapolis	Seattle
Dulles*	New Orleans	Tampa
Honolulu	Newark*	Washington National* (includes Andrews AFB, MD)

Pilot and Avionics Requirements

As established in FARs 91.131 and 91.215, the pilot and equipment requirements to operate in a Class B airspace are listed below.

Pilot Requirements

- To *land* or *take off* at any of the asterisked airports listed above, the pilot in command must hold at least a private pilot certificate.

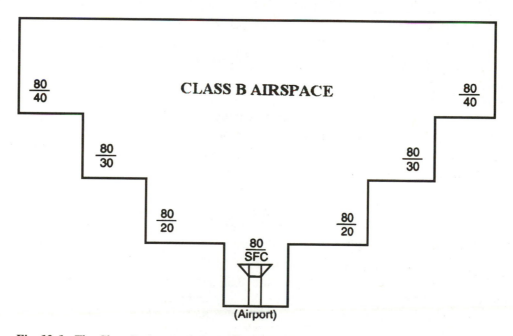

Fig. 12-1. *The Class B airspace is typically described as an upside-down wedding cake.*

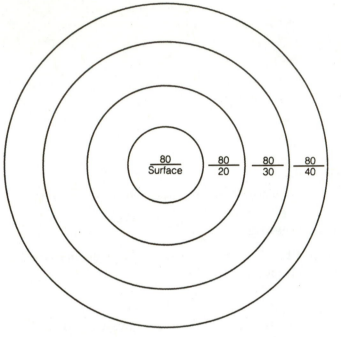

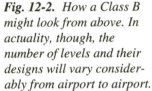

Fig. 12-2. How a Class B might look from above. In actuality, though, the number of levels and their designs will vary considerably from airport to airport.

- A student pilot may land, take off, or operate solo at any of the *nonasterisked* airports or within the related Class B airspace if he or she has received ground and flight instruction for that specific airport within the 90 days preceding the solo flight and has received a log book endorsement to that effect.

Avionics requirements

- For IFR operations, the aircraft must be equipped with an operable VOR or TACAN (Tactical Air Navigation military system).
- For both IFR and VFR operations. the aircraft must be equipped with
  1. An operable two-way radio with frequencies appropriate for the Class B airspace.
  2. An operable Mode 3/A4096-code transponder (or Mode S transponder) with Mode C altitude-reporting capability. This equipment is also required in the airspace within the 30-nautical-mile Mode C veil, other than the exceptions to be noted in a moment.

Participation: Voluntary or Mandatory?

When entering, departing, or transiting a Class B airspace, either VFR or IFR, you have only one choice. Participation (meaning establishing and maintaining contact with the controlling agency) is *mandatory.* FAR 91.131 makes this very clear:

> No person may operate an aircraft within a Class B airspace except in compliance with Para. 91-129 and the following rules.

A list of rules follows, including those affecting pilot qualifications and equipment requirements that I summarized above.

Identifying a Class B Airspace

Class B airspaces are easy to spot on the sectional charts because their locations are always outlined by a heavy blue square that boxes in the entire area. Figure 12-3, the Kansas City Class B airspace, illustrates the square as well as other features common to most Class B airspaces. I should stress again, though, that all Class B airspaces are hardly the same. Their basic structures are similar, but the individual designs, the number of levels, the height of the floors, or the airspace's vertical limits can vary considerably.

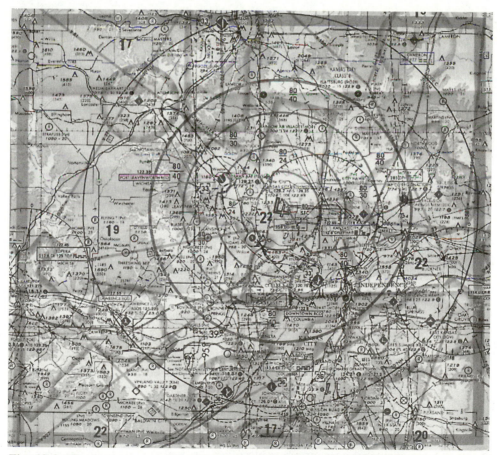

Fig. 12-3. The large square (blue on the sectional) identifies the existence of a Class B airspace within its borders, while the circles outline the levels, or shelves, of the airspace itself. The thin outermost circle represents the 30-mile veil radius, within which, barring certain exceptions, Mode C transponders are required.

The actual Class B airspace in Fig. 12-3 is not defined by the square or rectangle itself but by the solid blue circles or the irregular blue designs that surround the primary airport. The innermost circle represents the core of the airspace. In addition to the airport runway layout, certain key airport information, such as radio frequencies, field elevation, and length of the longest runway, are generally found in this area. There's another bit of information, too: Referring to Fig. 12-3 again, just to the right of the airport runway symbol you'll see "80/SFC." This means that the airspace core rises from the airport surface to 8,000 msl, which is then the common ceiling of the entire airspace. Each succeeding outward circle establishes the next layer or shelf, along with its ceiling and floor, as 80/24, 80/30, and so on. (These figures should, of course, be read in thousands of feet: 8,000/2,400, and so on.)

The Mode C Veil

We're back to this peculiarity of the Class B airspace and the area in which Mode C transponders are required. The thin blue circle, approximately 30 nautical miles out from the primary airport, defines the limits of the veil. Everywhere within that 30-mile radius, all aircraft, barring the exceptions I'll note in a moment, must have an operable Mode C transponder and the transponder must be tuned to the ALT position.

As to the exceptions: When the Class B airspace was initially proposed, along with the veil concept, pilots and FBOs who operated primarily at the small Class G or E airports out near the 30-nautical-mile boundry rose up almost en masse. They argued that it was totally unnecessary to require transponders 25 to 30 miles away from the Class B airport and with the floor of the overriding Class B extension usually four or five thousand feet above those peripheral airports. "Plenty of space to operate," they said, in essence. Furthermore, they added that the ruling would be costly to many aircraft owners who would either have to buy transponders in the first place or go to the expense of adding the altitude-reporting capability.

Give credit to those at the FAA: They listened to the arguments. They didn't do away with the veil, but they did establish some exceptions to the ruling which were intended to expire by December 30, 1993. As of the time of writing, however, SFAR (Special Federal Aviation Regulation) No. 62 still lives, as do the exceptions.

What SFAR 62 does is exempt aircraft that operate out of specified airports that are located on or within just a few miles of the 30-mile outer veil limit. More exactly, the SFAR exempts these aircraft "in the airspace at or below the specified altitude (I'll clarify that momentarily) and within a 2-nautical-mile radius , or, if directed by ATC, within a 5-nautical-mile radius, of an airport listed in…this SFAR."

The ruling goes on to say, in effect, that traffic from these airports to the veil boundry and out of the Bravo airspace must be as direct as possible, "consistent with established traffic patterns, noise abatement procedures, and safety." In other words, within an airport's 2-mile radius, no Mode C is necessary. Departing to the outer world, however, must be direct and within the altitude limits established for that particular Class B airspace. No flying around from airport to airport in this veil area without Mode C. You either stay local (within the 2-nautical-mile radius) or head straight out of the area. And the same applies when entering the veil to go to one of the exempt airports. As for touch-and-gos at the exempt airport, no problem. That's one of the reasons for the 2-mile (or it might be 5-mile) Mode C-free radius.

Now as to the "specified altitude" restriction. SFAR No. 62 lists the Class B airspaces and their nearby exempt airports. At the same time, the listing establishes the maximum altitude for each airport at which non-Mode C flight can be conducted. These altitudes vary from one Class B airport to another, but they are all consistent within each Class B's veil. For example, the altitude for exempt airports in the Atlanta veil is 1500 feet agl; in Detroit, it's 1400 feet, 2000 feet at Kennedy, 1500 at Tampa, and so on.

All told, there are about 300 airports across the country that qualify for this non-Mode C requirement. Consequently, if you're planning a trip to one that is in the outer portion of that 30-mile blue circle, you might check on SFAR No. 62. Found either at the beginning or the end of FAR Part 91, this will give you a complete listing of those excluded airports and their maximum operating altitudes at each of the Class B terminal areas.

In discussions with FAA officials as late as March 1998, it appears that the principal ingredients of SFAR 62 will remain, but under some new designation. Until there is further FAA action, however, FAR 91.215, now on the books, negates or contradicts SFAR 62. Although the FAR still establishes certain operating restrictions, it exempts *only* balloons and gliders from the Mode C requirement, along with powered aircraft that were originally manufactured without an electrical system and in which no such system has been subsequently installed. Otherwise, 91.215 states that *all* aircraft must be equipped "with an operable coded radar beacon transponder . . . having a Mode C capability." Until further notice, then, and despite the conflict between SFAR 62 and FAR 91.215, the rulings stated in the SFAR continue to live—which is a good thing. The rulings make sense and are certainly advantageous to pilots who are based at or fly regularly into these so-called fringe airports.

Beyond the veil...

While the focus so far has been on the sectional chart as the principal identification source for Class B airspaces, don't overlook the *A/FD*. As Fig. 12-4 illustrates, the *A/FD* provides a wealth of information about each airport, plus, in the case of Fig. 12-4, that Kansas City is a Class B airspace. Before going into any airport, whatever its classification, be sure to check the applicable *A/FD*. Only there, in conjunction with the appropriate Class B airports Terminal Area Chart (TAC), will you find the crucial data that you'll need, both on the ground and in the air.

Among its many features, the TAC explodes the area within the large blue square on the sectional and provides a more detailed "picture" of landmarks, reporting points, Approach Control frequencies, and other important surface features. Just be sure that it, like the sectional, is a current issue.

Entering the Class B Airspace

As previously indicated, I'll give examples of the radio calls themselves in the next chapter. Right now, let's just establish the rules, with the assumption, of course, that you have met the FAR pilot qualifications for Class B operations summarized earlier.

Using the Terminal Area Chart as the primary reference, always contact Approach Control on the frequency printed on the TAC well before you near the Class B shelf that you intend to enter. This means that the call should be made when you're about 15 miles

---

KANSAS CITY INTL (MCI) 15 NW UTC−6(−5DT) N39°17.86' W94°42.84'   **KANSAS CITY**
  1026   B   FUEL 100LL. JET A   LRA   .ARFF Index C   **H−2E, 4G, L−6H, A**
  RWY 01L−19R: H10801X150 (CONC−GRVD)   S−100, D−200, DT−350   HIRL CL   **IAP**
    RWY 01L: MALSR. TDZL. 0.3% down.   RWY 19R: ALSF2. TDZL. Rgt tfc.
  RWY 01R−19L: H95001X150 (CONC−GRVD)   D−200, DT−400   HIRL CL
    RWY 01R: ALSF2.   RWY 19L: MALSR. TDZL.   0.6% up.
  RWY 09−27: H9500X150 (ASPH−GRVD)   S−75, D−125, DT−180   HIRL CL
    RWY 09: MALSR. TDZL.   RWY 27: MALSR. VASI(V4L)—GA 3.0° TCH 58'. Rgt tfc.
  AIRPORT REMARKS: Attended continuously. Waterfowl on and in vicinity of arpt Oct 1−Dec 15 and Apr 1−May 30. 75'
    lighted obstruction located 1000' east and 1500' north of AER 19R indefinitely. When using high−speed exits
    C3, C5, C6, continue until first parallel taxiway, then use extreme caution when turning in excess of 90°. Prior
    approval required to park at airline gate areas. When using hi−speed exits C3 and C6 continue until first parallel
    twy and use extreme care when turning in excess of 90 degrees. Maximum gross weight limiting for Taxiway B8
    and postal apron is S−45,000 and D−60,000. Rwy 09−27 B747 max gross weight of 460000 lbs, L1011
    260000 lbs, DC−10−30 315000 lbs. Landing Fee. Flight Notification Service (ADCUS) available. NOTE: See Land
    and Hold Short Operations Section.
  WEATHER DATA SOURCES: ASOS (816) 243−6415. LLWAS.
  COMMUNICATIONS: ATIS 128.35 (816) 464−2762   UNICOM 122.95
    COLUMBIA FSS (COU) TF 1−800−WX−BRIEF. NOTAM FILE MCI.
    RCO 122.65 122.1R 112.6T (COLUMBIA FSS)
  ® APP/DEP CON 132.95 (010°−190°) 124.7 (191°−009°)
    INTERNATIONAL TOWER 128.2 125.75   GND CON 121.8 121.65   CLNC DEL 135.7 ◄
  AIRSPACE: CLASS B See VFR Terminal Area Chart.
► RADIO AIDS TO NAVIGATION: NOTAM FILE MKC.
    (H) VORTAC 112.6   MKC   Chan 73   N39°16.76' W94°35.48'   273° 5.8 NM to fld. 1060/8E.   HIWAS.
    DOTTE NDB (MHW/LOM) 359   DO   N39°13.25' W94°45.00'   015° 4.9 NM to fld. NOTAM FILE MCI.
    HUGGY NDB (LOM) 416   RN   N39°18.12' W94°51.07'   087° 6.4 NM to fld. Unmonitored.
    ILS/DME 109.7   I−RNI   Chan 34   Rwy 09   LOM HUGGY NDB. LOM unmonitored.
    ILS/DME 110.75   I−PVL   Chan 44(Y)   Rwy 01R
    ILS 110.5   I−DOT   Rwy 01L   LOM DOTTE NDB.
    ILS 109.1   I−PAJ   Rwy 19R
    ILS 109.55   I−DYH Rwy 19L

Fig. 12-4. *This is an example of how the A/FD identifies a Class B airspace and the Clearance Delivery frequency. (See arrows.)*

out from that shelf, as illustrated in Fig. 12-5. Otherwise, on a busy day with a steady flow of radio communications, you might be about to enter the airspace before Approach could respond to your call. And entry without a clearance is a distinct no-no.

Once Approach has cleared you, you'll be given headings and altitudes to fly until Approach authorizes you to change to the tower frequency. You're then in, or almost in, the airspace controlled by the tower, and the tower will give you the final landing instructions.

Let's set up another situation: The primary airport has sufficient traffic to warrant a Class B airspace, but you want to land at another field that lies under that airspace but outside of the primary airport's immediate vicinity. St. Louis is a good example. The core of the airspace from the surface to 8,000 feet surrounds Lambert Field. About 20 miles west-southwest is Spirit of St. Louis, a tower-controlled Class D airport. From a profile view, looking north, the Class B airspace appears as depicted in Fig. 12-6. You're approaching Spirit of St. Louis Airport from the west. Are you required to contact Approach? Yes, if your position and altitude would put you into any one of the floor/ceiling levels. No, if you stay below the floors. You must, however, contact this airport's control tower before you enter its airspace.

So much for the basic procedures for entering and landing in a Class B airspace. How about departing? Here are three situations:

In the first instance, you're leaving the primary Class B airport. Before doing so, though, you must receive a clearance from what is called "Clearance Delivery." Following the subsequent standard contacts with Ground Control and the tower, you then, in accordance with instructions you've received, call Departure Control after takeoff.

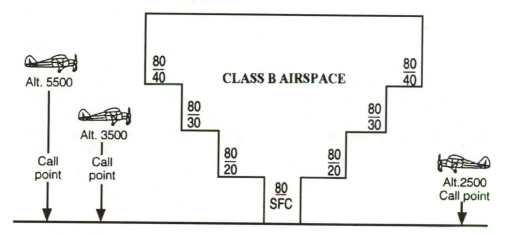

Fig. 12-5. *Hardly to scale, the purpose of this figure is to stress the need to contact Approach Control for clearance into one of the Class B levels 10 to 20 miles out from the level, or shelf, you intend to enter.*

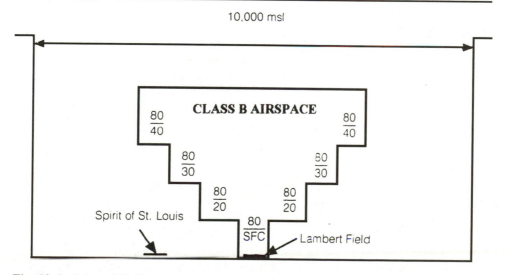

Fig. 12-6. *Spirit of St. Louis is an example of an airport that lies under, but is not in, the Class B airspace.*

All Class B and C airports, as well as many Class Ds, have this Clearance Delivery service. Operating on its own frequency, one of its purposes is to coordinate and communicate departure clearances in order to reduce congestion on the Ground Control frequency. If you'll refer back to Fig. 12-4, you'll note that its availability and frequency, where the service exists, are included in the "Communications" section of the *A/FD*.

In the second instance, you're leaving a secondary airport, such as Spirit of St. Louis, that lies under, not in, a Class B airspace. You want to go east and cruise at 5,500 feet, but since that altitude would quickly put you in the Class B airspace, you must contact Approach Control before penetrating any level of the airspace.

A question at this point: Why Approach and not Departure? Simply because, in this case, you're leaving a secondary field and will be *entering,* thus *approaching,* the airspace, and will be under the jurisdiction of the controllers handling all approaching aircraft. Conversely, if you're taking off from the primary Class B airport, such as Lambert Field, you're already in the core of the airspace, so Departure is the responsible controlling agency.

In the third instance, you're leaving from a secondary airport, like Spirit, and plan to fly locally or leave the area by staying well below all overriding Class B floors. Do you need to call Approach? No, because you'll be beneath, not in, the primary airport's protective airspace. You merely establish the normal contact with Spirit's tower and stay tuned to that frequency at least until you're out of the 5-mile Class D airspace.

One more situation: You're on a VFR cross-country flight and your desired route of flight takes you directly through a Class B airspace. You don't want to land anywhere in that airspace, but to go around the whole area would only add time and cost to the flight. Also, the ceilings are such that you can't climb above the top of the airspace. You must go through it or go around it. If you're not sure of your radio procedures, you'll probably opt for the latter and head out on a fair-sized detour, meanwhile watching the fuel dwindle and the bill go up.

Let it be said, however, that Class B airspaces can be transited if you meet the pilot and aircraft equipment requirements *and* if permission to enter the airspace has been granted. All it takes is a call to Approach approximately 15 miles out from the airspace floor you intend to enter. Give the usual IPAI/DS information, concluding with "Request clearance through the Class B airspace." After clearing you into the airspace, Approach takes it from there, with whatever vectors or directives are necessary, and notifies you when radar service is terminated once you're on the other side and clear of the area.

More on Class B Operations

A couple of other points about operating in or near Class B airspaces should be raised. First is the matter of cloud avoidance.

For those of us flying VFR, the general regulation relative to cloud avoidance is the 500-1000-2000 rule, meaning that you must be at least 500 feet below the clouds, 1000 feet above them, or 2000 feet from them horizontally. It's a different story when in a Class B airspace, though. There the rule is simply to *remain clear of clouds.* The reason is to minimize disruption of the flow of other traffic within the airspace. It's the pilot's job to adhere to the altitudes and headings assigned by Approach Control, but it's also the

VFR pilot's responsibility to remain VFR at all times. To do so under the 500-1000-2000 regulation could require altitude or heading changes in the effort to dodge clouds or cloud layers that loomed ahead. Depending on the volume of traffic in the airspace, these various course deviations by VFR aircraft (just to stay legal) could play havoc with Approach's ability to maintain the necessary sequencing and separation of aircraft in the airspace. Hence we have this clear-of-clouds ruling, which became effective with reclassification back in September 1993.

A second point: If you're flying outside a Class B airspace but under one of the floors, be sure that you have the current airport altimeter setting and then stay at least 200 feet below the floor in smooth air. If the air is rough, 400 to 500 feet below is better. You don't want to find yourself suddenly bounced up or caught in an updraft that would thrust you into the shelf and thus the airspace—which could easily happen if you were skimming along only a few feet beneath the floor.

Third: Just as you can't enter a Class B airspace without prior approval, IFR aircraft can't operate below one of the shelves, whether arriving or departing. You thus should have no concern about seeing a 747 pop up in front of you, even in a busy airline hub environment. What's good for the goose....

Fourth: You want to go through a Class B airspace but not land, so you contact Approach for entrance clearance and radar vectoring. In response, the controller denies the request—which is his or her right whenever a VFR aircraft is involved. It's the controller's responsibility to sequence and separate IFR aircraft, but if the weather is questionable or the volume of traffic heavy, the only choice might be rejection of your request. The controller just doesn't have time to handle a VFR operation under the circumstances.

Should denial be the response, don't argue, don't complain. The controller has reasons and is under no obligation to explain those reasons. The controller's word is final, and your alternative is cut and dried: Stay out and go around the whole airspace, even if it costs you time and fuel.

Further to that point, the weather might be fine and all the rest, but the controller can still deny entrance if the pilot comes across as uncertain, confused, or unknowing. These folks on the ground are busy enough without having to deal with incompetence, and how you sound over the air could lead them to that conclusion about you—false though it may be.

To reject a VFR pilot's request for the last reason is perhaps rare, but it does happen. Conversely, most controllers do everything they can to help all pilots—VFR or otherwise. After all, many are pilots themselves, and they have no trouble putting themselves in the left seat.

CLASS C AIRSPACE

Between the regulatory controls of the Class B and the tower-only Class D airspaces comes the Class C airspaces. Here there are controls, to be sure, but getting into or out of a Class C airspace is easier, the structure is smaller, and the traffic volume possibly less than in the Class B airspace. (The word "possibly" is used advisedly because some of these smaller airports, including Class D airports, could actually record more takeoffs and landings than a Class B airport. Much of that traffic, however, could consist of small, general aviation aircraft operating primarily under VFR conditions and thus not qualify for Class B or perhaps even C consideration.)

Relative to that, for Class C consideration the FAA has stipulated that the primary airport must have a control tower and meet at least one of these conditions:

- A minimum of 250,000 passenger enplanements a year *or*
- At least 75,000 instrument operations a year *or*
- At least 100,000 instrument operations a year at the primary and secondary airports included in the proposed Class C airspace

Identifying the Class C Airspace and Its Structure

A Class C airspace is easily identified on the sectional chart by two heavy magenta circles that surround the primary airport, as shown in Fig. 12-7 with that of the Columbia, South Carolina, airspace. For the most part, these are concentric circles, but, like the Class B airspaces, there could well be jogs or irregularities for any number of geographic or operational reasons. As with the Class B airspaces, the *A/FD* (Fig. 12-8) also identifies the airspace and provides much more information about the airport itself.

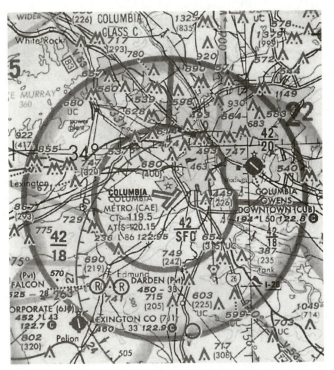

Fig. 12-7. *The solid magenta circular design identifies a Class C airspace.*

COLUMBIA
COLUMBIA METROPOLITAN   (CAE)   5 SW   UTC−5(−4DT)   N33°56.33' W81°07.17'   CHARLOTTE
   236   B   S4 . **FUEL** 100LL, JET A   OX 1, 3   TPA—See Remarks   LRA   ARFF Index C   H−4I, 6F, L−20F, 27B
RWY 11−29: H8602X150 (ASPH−GRVD)   S−72, D−200, DT−328, DDT−700   HIRL   CL   IAP
   RWY 11: ALSF2. TDZL. Trees.   **RWY 29:** MALSR. 0.5% up.
RWY 05−23: H8000X150 (ASPH−CONC−GRVD)   S−72, D−200, DT−315   HIRL
   RWY 05: 0.4% down.   **RWY 23:** VASI(V4L)—GA 3.0° TCH 54'. Thld dsplcd 1000'. Tree.
RUNWAY DECLARED DISTANCE INFORMATION
   RWY 05: TORA−8000   TODA−8000   ASDA−7000   LDA−7000
   RWY 23: TORA−8000   TODA−8000   ASDA−8000   LDA−7000
AIRPORT REMARKS: Attended continuously. PPR arpt manager before acft transporting explosives lands at fld. Obtain
   permission during business hours 1330−2200Z‡ Mon−Fri except holidays 803−822−5000. All 180°turns on
   grooved surfaces prohibited. Fee for commercial aircraft over 15,000 pounds. Bird activity on and in vicinity of
   arpt. Opr of ultralight vehicles prohibited. TPA for propellar acft 1236(1000); TPA for turbprop 2036(1800).
   Sports complex with numerous flood lgts approximately 6500' from apch end Rwy 11. NOTE: See Land and Hold
   Short Operations Section.
WEATHER DATA SOURCES: ASOS (803) 822−4168
COMMUNICATIONS: ATIS 120.15   **UNICOM** 122.95
   ANDERSON FSS (AND) TF 1−800−WX−BRIEF. NOTAM FILE CAE.
   RCO 122.1R 114.7T (ANDERSON FSS)
 Ⓡ **APP/DEP CON** 124.15 (110°−289°) 133.4 (290°−109°)
   TOWER 119.5   **GND CON** 121.9   **CLNC DEL** 119.75
AIRSPACE: CLASS C svc continuous ctc **APP CON**
RADIO AIDS TO NAVIGATION: NOTAM FILE CAE.
   (H) VORTAC 114.7   CAE   Chan 94   N33°51.44' W81°03.23'   328° 5.9 NM to fld. 410/02W.
   MURRY NDB (LOM) 362   CA   N33°58.03' W81°14.69'   110° 6.5 NM to fld.
   ILS 110.3   I−CAE   Rwy 11.   LOM MURRY NDB.
   ILS 108.3   I−VYK   Rwy 29. LOC unusable .15 NM to rwy thld.
   ASR

Fig. 12-8. *What the A/FD tells about a Class C airspace and the airport itself.*

Another helpful identifying feature in some sectionals is inclusion of a wide blue square, similar to that for the Class B airspaces, that encompasses all of the Class C area, and more. The boxed-in area is enlarged elsewhere on the chart and provides more geographic detail than is otherwise possible. This doesn't apply to all Class C airspaces, however. For example, the Cincinnati sectional enlarges the Cincinnati and Dayton areas but not Columbus, which is also a Class C airspace. And the same is true of many other Class C airspaces. The feature is helpful but not essential to the easy identification of the airspace.

Structurally, the Class C is simple. It has only two rings, or circles, with the inner circle rising from the surface to approximately 4,000 feet agl above the airport. The radius of this circle extends 5 nautical miles from the airport, while the second, or outer, circle has a 10-nautical-mile radius, perhaps with varying floor levels but with the same common ceiling of the inner circle.

So far, as Fig. 12-9 illustrates, the shape of a Class C is rather like a Class B airspace, except that it is considerably smaller in radius. One design feature, however, makes the Class C airspace unique: the *outer area*. This is an area that begins at the edge of the outer circle and extends another 10 nautical miles, giving the entire space, in effect, a 20-nautical-mile radius. Vertically, the outer area rises from the lower limits of radio and radar coverage up to altitudes of 10,000 to 12,000 feet agl, or, as AIM explains: "...up to the ceiling of the Approach Control's delegated airspace...."

Although Fig. 12-9 does show the typical Class C configuration, go back to Fig. 12-7 momentarily. You'll notice that the outer area is not depicted on the sectional. It's still there, however, so be aware of its existence. What it means in terms of pilot operations I'll review shortly.

Class C Pilot and Equipment Requirements

Class Cs have no special pilot requirements; students on up may enter and fly in the airspace. The only equipment needed is a two-way radio capable of sending and receiving on the appropriate Class C frequencies, plus a Mode C for operations within the airspace and above it up to and including 10,000 feet msl.

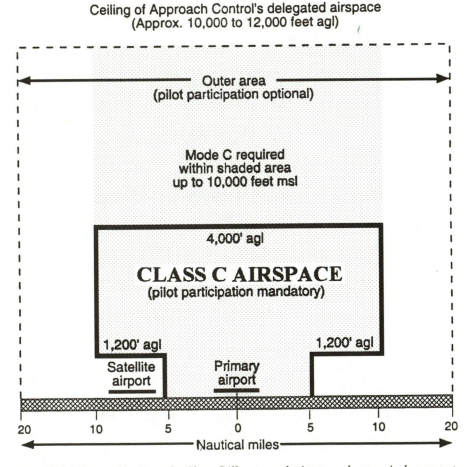

Fig. 12-9. *The profile view of a Class C illustrates the inner and outer circle concept, plus the outer area, which is all of the airspace except that lying within the heavy lines.*

Pilot and Controller Responsibilities in a Class C Airspace

From the pilot's point of view, and according to FAA regulations, no literal "clearance" into a Class C (as when entering a Class B airspace) is required. As FAR 91.130 states it:

Each person must *establish* (italics mine) two-way radio communications with the ATC facility providing air traffic services prior to entering that airspace and thereafter maintain those communications while within that airspace.

While this refers primarily to arriving and through flights *before* they enter the outer circle (not the outer area), the same basic principle applies to departing aircraft. In both respects, there is little difference, up until now, in radio procedures between a Class B and Class C airspace. What is different is the meaning of the word *establish*. The meaning behind the ruling is this: If you call a Class C Approach Control and the controller replies with your call sign such as "Cherokee One Four Six One Tango, stand by," you have now *established* "two-way radio communications with ATC…" and can enter the airspace. But, if ATC's response to your introductory call is "Aircraft calling Columbia Approach, stand by," you have *not* established two-way radio communications and you are *not* authorized to enter the Class C area. The key is the inclusion of your call sign in ATC's response. If it's there, you can continue right on into the airspace. Otherwise, stay outside "until further notice."

The same radio communication requirement applies to departing aircraft, but how that's done when leaving the primary airport is for later discussion. At the same time, aircraft departing a satellite airport within the inner circle must establish radio contact with Departure Control as soon as possible after takeoff, and they must maintain that contact while in the Class C airspace.

Once radio contact has been established, ATC will provide the following services within the 10-nautical-mile radius:

- Sequencing of all aircraft arriving at the primary airport
- Separating VFR aircraft from IFR aircraft
- Providing traffic advisories and conflict resolution
- Issuing safety alerts, if necessary, to VFR aircraft

The last point warrants clarification for those operating VFR. Unlike the Class B airspace, ATC is not required to *separate* VFR aircraft in the Class C, but may do so in certain locations if the density of traffic indicates the need. Separation involves whatever vectoring or altitude changes are necessary to maintain a given distance between aircraft. *Advisories* alert the pilot to bearings, approximate distance, and altitude in relation to other aircraft that might pose a potential safety problem. It is then the pilot's responsibility to scan the skies, locate the traffic, and advise ATC when the traffic has been spotted.

A *safety alert* is issued when conflict between two aircraft, or an aircraft and a ground obstruction, seems imminent. When transmitted, it means that the receiving pilot should change course, altitude, or possibly both, *now*. The situation is reaching emergency proportions. Alerts will be few, however, if advisories are heeded and the pilot is attentive to what's going on outside the cockpit.

Chapter Twelve

The Outer Area

The outer area is not regulated airspace, not really part of the official Class C airspace; therefore two-way radio communication in this area is not mandatory. Although controllers are learning to live with it, the outer area has been a source of some dispute. It can be both an advantage and a disadvantage for pilots, while at the same time placing an added workload on the controller. To explain:

Let's say that you leave the primary airport and go to the outer area to practice maneuvers. You've maintained radio contact with Approach (actually Departure Control) and while you're in the outer area, the controller must provide the advisories and alert services just as though you were in the inner or outer circles. But, as soon as you hit that outer area, you have the right to request that those services be terminated. Unless you do, the controller has no alternative but to continue them.

Another example: You leave a satellite airport in the outer area and intend to stay in that area and not enter the Class C airspace. If you want the advisory and alert services, you must specifically request them. Otherwise, you won't hear from Approach at all.

Said simply, the VFR pilot is in the driver's seat. He can request, not request, or terminate the services in the outer area at will. The controller has no such prerogative. The potential—or real—problem lies in the fact that a whole lot of aircraft could be milling around and requesting radar service in the outer area, which extends up to about 10,000 feet. ATC, having to provide that service to these aircraft, could be so busy that it would be forced to temporarily deny landing or transiting aircraft entry into the airspace.

The policy states that clearance into the airspace is not required, but some controllers have found that they have had to tell pilots who are at the fringes of the outer circle (not the outer area) to "remain clear." The volume of activity in the outer area has, at times, created such an additional workload that service to those wanting to enter the area has been delayed. Consequently, the policy can benefit one group and adversely affect another—with the controller caught in the middle.

Not debating the value of radar service or your right to receive it, it would be considerate of others to terminate the service if you're going to fly in and around the outer area. If there's a valid reason for wanting the service, fine, but too many pilots who require advisories only consume airtime, add work for the controller, and possibly delay others' entry into the airspace.

So, let's summarize the Class C airspace concept this way:

- The only equipment required is a two-way radio capable of communicating with ATC and a Mode C transponder for operations within the Class C airspace and above it, up to and including 10,000 feet msl.

- Radio contact with ATC must be established before entering the airspace.

- ATC assumes that Class C service is wanted as soon as radio contact is established. If you call Approach and the controller responds to you with "Cherokee One Four Six One Tango, stand by," you have satisfied the communications requirements and may enter the airspace—unless you are told to remain clear.

- Participating aircraft must comply with all ATC vectors, instructions, and other directions.

- ATC will provide standard separation between IFR aircraft (1,000 feet vertically, 3 miles horizontally); VFR-IFR separation of 500 feet vertically, advisories and, if necessary, safety alerts to VFR aircraft. No separation is normally provided between VFR aircraft.

- ATC will sequence landing aircraft in daisy-chain fashion and turn each aircraft over to the local Control Tower (primary or secondary airport) for final landing instructions.

- Outer area services are provided as long as radio communications are maintained. They are terminated only when the pilot advises that Class C services are not desired.

- Class C service to aircraft landing at satellite airports within the Class C airspace is discontinued when the pilot is instructed to contact that airport's Control Tower.

- Aircraft departing a satellite airport will not receive Class C service until they have established radio communications and are radar identified.

- Participation is mandatory for any aircraft landing, departing, or transiting the 10-mile Class C radius (except under the floor of the outer circle). It is voluntary if landing, departing, or operating within the outer area.

Conclusion

That's the story of the Class C airspaces—the sort of junior Class Bs in the airspace system. There are currently 122 such Class Cs around the country, including Alaska, Hawaii, San Juan, and a number of Air Force bases, so you're more than likely to encounter one if you do any flying beyond the Class D or nontower airports. It's thus rather important to know what they are, their structure, and their requirements.

One problem, however, accompanies operating in an environment such as a Class B or C airspace. It's the problem of complacency. You're under radar scrutiny and a controller is giving you headings and altitudes to fly as well as traffic advisories. With that extent of ground guidance, the tendency is strong to keep one's head in the cockpit rather than scanning the skies to see what's going on outside.

As *AIM* puts it, in a rather lengthy sentence, in reference to Class C airspaces:

This program is not to be interpreted as relieving pilots of their responsibilities to see and avoid other traffic operating in basic VFR weather conditions, to adjust their operations and flight path as necessary to preclude serious wake encounters, to maintain appropriate terrain and obstruction clearance, or to remain in weather conditions equal to or better than the minimums required by FAR 91.155 (the basic VFR weather minimums).

AIM concludes, as I've stated before, that if a given ATC-directed heading or altitude is likely to compromise your responsibility with respect to other traffic, terrain, vortex exposure, or weather minimums, you should advise the appropriate ATC and obtain alternate instructions.

Said simply, don't be lulled into complacency just because a controller at a radar screen is telling you what to do. Keep your ears tuned to his or her instructions, but above all, keep your eyes open for any condition that might become an unwelcome interruption of your flight. That's your job as a VFR pilot in these areas of controlled and sometimes voluminous traffic.

THE TRSA—A FEW WORDS ABOUT THE TERMINAL RADAR SERVICE AREA

This element of the original terminal radar program is a peculiar mix of airspaces with some rather unique features. For example: (1) The TRSA has never been a controlled airspace from a regulatory aspect; (2) it has no counterpart in the ICAO airspace system, and is thus identified by a name rather than a letter; (3) because of its level of IFR operations and passenger enplanements, it lies between a Class C and a Class D airport but emerges as neither; (4) it provides radar approach control service to IFR and VFR aircraft, but VFR traffic has the option of using or not using the service; (5) there are only a relatively few TRSAs around, and sooner or later all will probably either move up to a Class C or drop down to a Class D tower-only airport; and (6) in terms of a mix, you might view the TRSA as a Class D towered airport surrounded by circles or segments of partially controlled airspaces.

Despite their relative rarity, plus their peculiarities, I can't overlook their existence here or the procedures involved in entering, departing, or transiting one of these neither-nor entities.

Identifying TRSAs

This is easy to do. They're spotted on the sectional chart by the black (almost gray) bands that surround the primary airport. The bands at the Binghamton, New York, TRSA (Fig. 12-10) reflect a basically circular pattern, which is generally, but not necessarily, the typical design. For instance, some TRSAs have odd-shaped extensions for IFR-aircraft protection; others have cutouts similar to the one around the Tri-Cities Airport in Fig. 12-10; some have three or four bands with varying floor levels; and there are those that have only one perfectly circular ring or band. So design uniformity is not a TRSA trademark.

Along with what the sectional tells you, more information about a TRSA is available in the *A/FD*. Figure 12-11 illustrates the Binghamton Regional Airport data and (by the arrow) how the type of airspace is identified.

TRSA Services Provided

Note first, an important point for VFR pilots: Specific clearance into a TRSA by Approach Control is not required. If, for whatever reason, you don't want to use the available radar services, you can refuse them in your initial call to Approach or, if departing,

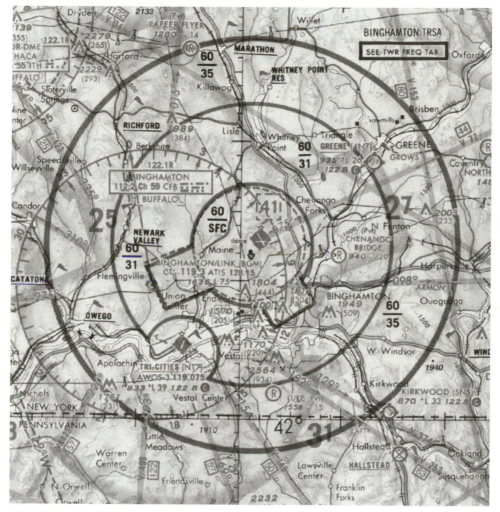

Fig. 12-10. *The TRSA structure closely resembles those of Class B or C airspaces.*

to ground control by merely stating, "Negative TRSA service." A question arises, however, as to why someone would not take advantage of the assistance ATC is ready to provide. Personally, I've always found it rather nice to have help through a perhaps busy terminal environment. But, should you choose to refuse it, just continue on course until you're within a few miles of the airport and then contact the tower for landing instructions. Meanwhile, though, it would be smart to keep monitoring the Approach frequency to learn what you can about other traffic in the area: where it is, its altitude, and its intentions.

On the other hand, if you have decided that you *do* want the TRSA services, make that known in your initial call. For example, when arriving and following the basic IPAI/DS, all you need to say is this:

```
BINGHAMTON REGIONAL/EDWIN A LINK FLD   (BGM)   7 N   UTC–5(–4DT)                          NEW YORK
  N42°12.51' W75°58.78'                                                                   H–3I, 6I, L–25A
  1636   B   S4   FUEL 100LL, JET A   OX 1, 2, 3, 4   TPA—2430(800)   ARFF Index B         IAP
  RWY 16–34: H7501X150(ASPH–GRVD)   S–98, D–125, DT–187   HIRL   0.9% up NW
    RWY 16: MALSR. PAPI(P2L)—GA 3.0°TCH 52'. Thld dsplcd 501'.
    RWY 34: MALSR. PAPI(P2L)—GA 3.0°TCH 44'. Thld dsplcd 300'.
  RWY 10–28: H4998X150 (ASPH–GRVD)   S–98, D–125, DT–187   MIRL
    RWY 10: REIL. VASI(V4L)—GA 3.0°TCH 55' (Unmonitored).        RWY 28: REIL. VASI(V4L)—GA 3.0°TCH 45'.
  RUNWAY DECLARED DISTANCE INFORMATION
    RWY 16:   TORA–7501   TODA–7501   ASDA–7001   LDA–6500
    RWY 34:   TORA–7501   TODA–7501   ASDA–7000   LDA–6700
  AIRPORT REMARKS: Attended continuously. Lgtd structure 118 ft AGL 2000 ft south of apch end Rwy 10. Ldg fee for
    acft over 4000 pounds; except when fuel purchased. PAPI Rwy 16 only on when rwy active; other times ctc
    tower. Bird activity on and in vicinity of arpt. PPR 12 hours for unscheduled air carrier ops with more than 30
    passenger seats 0400–1100Z‡ call arpt manager 607–763–4474. ARFF available 1100–0400Z‡ and/or until
    15 minutes after the last scheduled air carrier or commuter. Tower unable to see hard surface W of
    maintenance building. Fuel svc 1030–0400Z‡, Mon–Fri and 1100–0200Z‡ Sat–Sun and holidays; svc fee other
    times call 607–770–1093. No fixed base operator svcs avbl 0400–1000Z‡ Sun–Fri and 0230–1000Z‡ Sat
    without prior coordination; call 607–770–1093. Turbine engine powered acft operated under FAR 121.189 and
    135.379 are advised that clearance distances may not be adequate for takeoff when acft are holding on Twys
    B–C–D–F and J for Rwy 16–34 and Twy F for Rwy 10–28. NOTE: See Land and Hold Short Operations Section.
  WEATHER DATA SOURCES: ASOS (607) 729–8335.
  COMMUNICATIONS: ATIS 128.15
    BUFFALO FSS (BUF) TF 1–800–WX–BRIEF. NOTAM FILE BGM.
    RCO 122.1R 112.2T (BUFFALO FSS)
  Ⓡ APP/DEP CON 118.6 (185°–050°) 127.55 (051°–184°) (1100–0500Z‡)
  Ⓡ NEW YORK CENTER APP/DEP CON 132.175 (0500–1100Z‡)
    TOWER 119.3   GND CON 121.9   CLNC DEL 125.05
→ AIRSPACE: TRSA svc ctc APP CON
  RADIO AIDS TO NAVIGATION: NOTAM FILE BUF.
    (L) VORTAC 112.2   CFB   Chan 59   N42°09.45' W76°08.19'   076° 7.6 NM to fld. 1570/10W.
    SMITE NDB (LOM) 332   BG   N42°06.29' W75°53.47'   340° 7.4 NM to fld.
    ILS 110.3   I–BGM   Rwy 34.   LOM SMITE NDB.
    ILS 110.3   I–AAJ   Rwy 16.
```

Fig. 12-11. What the A/FD says about the Binghamton, NY, TRSA.

Binghamton Approach, Cherokee One Four Six One Tango over Cortland at five thousand five hundred, landing Binghamton (or Link), squawking one-two-zero-zero (or "squawking twelve" or "squawking VFR") with Echo. Request TRSA service.

And when departing:

Binghamton Clearance, Cherokee One Four Six One Tango at Miller Aviation, VFR northwest to Oswego at four thousand five hundred with Echo. Request TRSA service.

That's all there is to it. And, if you don't want the service, merely substitute the word "Negative" for "Request."

The TRSA service is designed to provide separation between all IFR and all participating VFR aircraft. VFR aircraft will be separated from other VFR/IFRs by 500 feet vertically, or by visual separation, or by target resolution, meaning that space is maintained on the radar screen between targets when using the broadband radar system. In other words, there is "green between" the targets on the screen.

For arriving traffic, Approach will provide these services until the aircraft nears the control tower's airspace, at which point Approach will tell the pilot to contact the tower on the tower's assigned frequency. From then on, the tower controller takes over until the aircraft is safely down and clear of the runway. Ground Control enters the picture at that point for taxi clearance to the pilot's desired ramp location.

When departing, the sequence of calls is just as in a Class B or C airspace:

1. ATIS
2. Clearance Delivery
3. Ground Control
4. Tower
5. Departure Control

Once at or just beyond the TRSA's outer limits, Departure will normally so advise participating VFR aircraft.

AIM, under the heading of "Terminal Radar Services for VFR AIrcraft," has more on the subject than in this brief overview. Consequently, should you be planning to fly into a TRSA area, you would do well to consult that resource for further details.

CONCLUSION

As some of the foregoing might be a little confusing, Fig. 12-12 summarizes the operating requirements of the various classes of airspace in the terminal environment. In only two situations is the VFR pilot required to contact Approach or Departure Control: when entering, departing, or transiting (1) a Class B airspace or (2) a Class C airspace. Going over, around, or under either airspace requires no radio contact (or, as the term goes, "participation"). Of course, it's another matter if an airport has an operating control tower, such as a Class D. Then approval to enter, leave, or transit that Delta airspace is always required.

I don't want to beat this next point to death, but safety demands repetition: Operating in a radar-controlled environment in no way relieves VFR pilots of the responsibility to maintain constant surveillance of the skies around them. It's too easy to be lulled into a false sense of security when someone on the ground is vectoring you, advising you of other traffic, and generally guiding or helping you through the airspace.

If there's one failure I've noticed among professional instructors, students, and licensed pilots, it's their propensity to keep their eyes either in the cockpit or straight ahead. They apparently haven't been taught to develop a swivel neck and maintain a constant scanning of as much of a 360 degree circle as aircraft structure and physical limitations permit. Back in the days of World War II, this point was continually emphasized in pilot training. The reason being, of course, that you could never be sure what unfriendly fellow was out there with murderous intent. The best insurance was to see him first, take evasive action if necessary, and then fire away.

Today, it isn't an accepted practice to fire at potential airborne targets, but seeing and evading are still not just acceptable practices, they are *life-saving*. If, in VFR conditions, you

	Pilot participation required	Two-way radio communications required	Mode C transponder required[2]	VOR required	Special pilot requirements	ATC clearance required to enter airspace	Sectional chart indication
Class B Airspace	✓	✓	✓[1]	✓	✓[2]	✓	Solid blue lines[3]
Class C Airspace	✓	✓	✓			4	Solid magenta lines
Class C Airspace Outer Area	5	5	5				Not shown
TRSA	6	6					Solid black lines
Class D Airspace	✓	✓				✓	Blue segmented lines/circle
Other radar or non-radar App/Dep control airports	6	6					Not shown; refer to *A/FD*

[1] See FAR 91.215 and SFAR No. 62 for regulations and veil airports.

[2] See requirements earlier in this chapter and FAR 91.131.

[3] Mode C limits shown by thin blue circle surrounding entire Class B airspace.

[4] See requirements earlier in this chapter.

[5] Required if radar service desired.

[6] Required if radar service desired. Otherwise, only in Class D 5nm airspace.

Fig. 12-12. *This chart may help clarify some of the airspaces' operating regulations and requirements.*

have your head in the cockpit, whatever the reason, for more than about 15 seconds, you ought to be getting a little nervous. That's too long, considering the rate of closure between even light aircraft. Flying in instrument conditions is one thing: attention to the instruments is essential. VFR is something else. "Head up and unlocked" should be the catch phrase of every pilot.

All of which has nothing to do with radio communications—except to suggest that in the process of communicating and being monitored on radar you should not allow yourself the luxury of relaxed alertness. Your safety in flight is enhanced by those on the ground who are advising you and perhaps directing you, but it's still your job to be your own protector and defender. A swivel neck and sharp eyes are the best shields against disaster in the sky.

13
Using Approach/Departure Control in the Controlled Airport Environment

As is hopefully apparent, the flow of this book, after the first opening chapters, basically began with radio communications and operations at the relatively simple multicom and unicom (Class G and E) airports. Then, in Chap. 11, we moved to the slightly more complex tower-controlled Class D airspace, followed in Chap. 12 by a discussion of the Class B (Bravo) and C (Charlie) even more strictly controlled airspaces. In other words, the sequence of subject matter was designed to progress from the "simple" to the "complex."

Particularly in Chaps. 11 and 12, I frequently referred to that ATC function called "Approach/Departure Control" but intentionally avoided discussing the details relative to its purpose, where it's located (on the airport property, I mean), when it should be contacted, or how communications with it should be conducted. Now, however, because of the important role it plays in the Class B, C, airspace operations and some Class D, understanding its function and what is expected of the pilot is essential. Of course, if you forever limit your flying to the Class G and E airports and airspaces, the importance of that understanding wanes. For those who want to enjoy the full benefits of flying, however, it's a different story. So, assuming the likelihood of broad rather than limited flying horizons, let's look more closely at this element of the ATC system.

(*Note*: To simplify wording, I'll generally refer to the function as "Approach," "Approach Control," or, if it won't be confusing, to "ATC"—"Air Traffic Control." The apparent emphasis on "Approach," however, in no way minimizes the role of "Departure" in the overall control of terminal area traffic.)

THE ROLE OF APPROACH AND DEPARTURE CONTROL

Described in one word, the Approach/Departure Control facility is a middleman. In this respect, Departure assumes radar and communications control of a departing aircraft after it has taken off from a tower-controlled airport and has been released by the tower controller. Departure Control then guides the aircraft through the 30- to 40-mile airspace for which it is responsible and out into the open country. At that point, it either turns the aircraft over to an Air Route Traffic Control Center or terminates radar contact. Conversely, Approach Control leads arriving aircraft from the outside world into the Class B or Class C airspace. When within 5 miles or so of the airport, it turns the aircraft over to the tower for landing instructions.

As is probably self-evident, the basic role of Approach/Departure is to ensure the orderly flow of traffic into and out of the busier terminals. Without it, the tower controller at higher-volume airports would face an impossible task of vectoring, sequencing, separating, and directing landing and departing aircraft—all within 30 miles or so of the primary as well as satellite airports. It just couldn't be done with any measure of safety. Hence the need for and value of this middleman.

WHERE IS APPROACH/DEPARTURE LOCATED?

What I mean by the question above is the physical location of the facility on the airport—not which airports have the service. The location is either in the tower structure of the primary airport or in a nearby radar-equipped building where it can coordinate closely with the tower personnel.

Three abbreviations commonly establish the locations: Tracab (terminal radar control in the tower cab); Tracon (terminal radar control); and Rapcon (radar approach control). All provide the same service, but Tracabs are relatively rare, if not extinct today, with Tracons the most common. The latter, usually located in the tower structure below the tower cab, is a darkened room, complete with radar, radio, telephone, and computer equipment. A Rapcon is associated with a military Approach/Departure facility that provides radar service to nearby civilian airports and civil aircraft.

To complete the picture, though, the service doesn't always require a facility to be on or near a given airport. Recall, for example, the Clarksburg, West Virginia, illustration in Chap. 11. There, when the local Approach/Departure is closed, the Cleveland Center handles arriving and departing traffic through its remoted radar and communications facilities.

Also to be noted is that fact that some airports have nonradar Approach Control. One example is Great Barrington, Massachusetts (Fig. 13-1), which is served by the Albany, New York Approach/Departure facility. In instances such as this, radar service might well be offered up to a certain distance or down to a certain above-ground-level altitude but not available thereafter. Whatever the reason for the limitation—distance, terrain, lack of instrument approach facilities at the particular airport—once that point or altitude is reached, Approach would resort to "manual" control via sequencing or timing until the aircraft reported itself down and clear of the runway. The FAA's *Air Traffic Control* Manual 7110.65J contains the procedures for nonradar approaches and departures, but these largely concern IFR operations and are thus beyond the purpose of this book. Be aware, however, that such services do exist.

The *A/FD* is the only source to determine whether a given airport has the service. If an ® precedes "APP/DEP CON," radar control is available. If there is no ®, the service is nonradar. And, if you don't see "APP/DEP CON" on the *A/FD* at all, the service doesn't exist.

DEPARTURE FROM A CLASS B AIRPORT

To be more specific about all of this, along with examples of the recommended radio phraseology, let's say that you're on the ramp at Kansas City International, which, as you know by now, is a Class B airport. The engine is started and you're ready to go.

First, tune to the ATIS and remember the information as well as the phonetic designation of the report. Now you make a call to a function that I've previously referred to

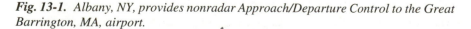

GREAT BARRINGTON   (GBR)   2 W   UTC−5(−4DT)   N42°11.05′ W73°24.19′   **NEW YORK**
   739   B   S4   **FUEL** 100LL   L−25C, 28H
RWY 11−29: H2579X50 (ASPH)   S−8   LIRL   IAP
   RWY 11: Thld dsplcd 170′. Trees.   **RWY 29:** VASI(NSTD)—GA 3.0° TCH 13′. Thld dsplcd 75′. Trees.
   AIRPORT REMARKS: Attended dalgt hours. Arpt lgts opr dusk−0400Z‡. ACTIVATE LIRL Rwy 11−29 and VASI Rwy 29
      after 0400Z‡—121.6. For rotating bcn after 0400Z‡ call 413−528−1010. Rwy 11 lgtd thld relocated 170 ft:
      2409 ft of rwy usable for ngt ops. Rwy 29 nstd VASI(VIL). Ngt ops only.
   COMMUNICATIONS: CTAF/UNICOM 122.8
    BURLINGTON FSS (BTV) TF 1−800−WX−BRIEF. NOTAM FILE BTV.
➤ **ALBANY APP/DEP CON** 125.0
   RADIO AIDS TO NAVIGATION: NOTAM FILE BTV.
    CHESTER (L) VOR/DME 115.1   CTR   Chan 98   N42°17.48′ W72°56.96′   266° 21.2 NM to fld. 1600/13W.
    NDB (MHW) 395   GBR   N42°10.98′ W73°24.24′   at fld. Unusable byd 20 NM.

Fig. 13-1. *Albany, NY, provides nonradar Approach/Departure Control to the Great Barrington, MA, airport.*

only in passing: Clearance Delivery (CD)—a function that exists primarily at Class B and C airports as well as some Class Ds.

The purpose of Clearance Delivery is to relieve congestion or excessive communication demands at the Ground Control position. CD has its own frequency, which is published in the *A/FD*. Once contacted, CD's job is to relay to the IFR pilot his IFR clearance. For those going VFR, the instructions consist of the following sequence:

1. Clearance to depart
2. Initial altitude to maintain
3. Departure Control frequency
4. Squawk code

As CD relays this data, be sure to copy it, and then, in an abbreviated form, read it back to the controller so that he or she knows that you have correctly recorded the instructions. (*Suggestion:* To make copying easier and faster, write down the main heading in advance and then merely fill in the blanks.)

After listening on the frequency for a few seconds to be certain that the air is clear, the dialogue would go like this:

You: International Clearance, Cherokee One Four Six One Tango.

CD: *Cherokee One Four Six One Tango, International Clearance.*

You: Clearance, Cherokee One Four Six One Tango at the general aviation ramp with India. VFR to Omaha, requesting six thousand five hundred.

CD: *Roger. Cherokee One Four Six One Tango cleared into the Bravo airspace. Maintain four thousand. Departure frequency will be one one niner point zero. Squawk two five two zero.*

You: Understand Cherokee One Four Six One Tango cleared into Bravo airspace, maintain four thousand, one one niner point zero, two five two zero. Six One Tango.

CD: *Readback correct, Six One Tango. Contact Ground.*

You: Will do. Six One Tango.

Two points: When Clearance Delivery gives you your clearance, that is your permission to operate your aircraft in the airspace. That was it, and no further authority is required. Second, you are given a transponder squawk—2520 in this case. Immediately set your transponder accordingly. Don't wait. You might forget either the code or to set it at all. And turn the transponder switch to STANDBY—not the ON or ALT position. Change to ALT only after you've been cleared for takeoff and are taxiing from the hold line to the active runway.

Now comes the call to Ground Control:

You: International Ground, Cherokee One Four Six One Tango at general aviation ramp with clearance and Information India. Ready to taxi.

> *CD:* *Cherokee One Four Six One Tango, taxi to Runway One Niner.*

You taxi out, complete the pre-takeoff check, contact the appropriate FSS to open your flight plan, and move to the runway hold line. Next comes the call to the tower:

> **You:** International Tower, Cherokee One Four Six One Tango ready for takeoff (or "ready to go").

> *Twr:* *Cherokee Six One Tango, Roger. Turn right to two seven zero. Cleared for takeoff.*

> **You:** Roger. Cherokee Six One Tango. (Now turn your transponder to ALT.]

After you're airborne, the tower contacts you again:

> *Twr:* *Cherokee Six One Tango, contact Departure.*

> **You:** Roger. Cherokee Six One Tango. Good day.

Note that the tower said nothing about the frequency on which to contact Departure. Also, you didn't have to request the frequency change. The reason is that all factors relative to your clearance had already been coordinated between Clearance, Tower, and Departure.

If you have two radios, change the frequency in one from Ground Control to Departure just before calling the tower. If you can set up your radios so that you're always one step ahead of the facility you're currently talking to, the transition from one to the next will be simplified. I discussed this before, but it's worth reemphasizing as just one more element of good cockpit organization.

Now make the next call following the tower's approval to contact Departure:

> **You:** Kansas City Departure, Cherokee One Four Six One Tango is with you, out of one thousand four hundred for four thousand. [Departure is well aware of your clearance and your squawk code. Consequently, all you need to do is identify your aircraft, your present altitude, and the altitude to which you are climbing. "With you" simply establishes the contact.]

> *Dep:* *Cherokee Six One Tango, radar contact. Climb and maintain six thousand five hundred. Turn right heading three four zero.*

> **You:** Roger. Out of one thousand seven hundred for six thousand five hundred. To three five zero. Cherokee Six One Tango.

> *Dep:* *Cherokee Six One Tango, that heading is three four zero.*

> **You:** Roger. Three four zero on the heading. Cherokee Six One Tango.

A few minutes later:

> *Dep:* *Cherokee Six One Tango, position two zero miles north of International, departing the Bravo airspace. Resume own navigation. Stand by for traffic advisories.* [This means that you are leaving the airspace but still in departure control's radar area. He thus wants you to stay with him for possible traffic advisories.]

You: Roger. Cherokee Six One Tango. Standing by. [You might not hear anything further from Departure until you leave the radar coverage area as follows.]

Dep: *Cherokee Six One Tango, radar service terminated. Squawk, one two zero zero, change to advisory frequency, approved.* ["Advisory frequency" is any frequency you now wish—Center, Flight Service, enroute airport towers, etc. There is, however, a chance that radar service will not be terminated and that Departure will "hand you off" to the appropriate Air Route Traffic Control Center for further traffic advisories. I'll get into that in the next chapter.]

You: Roger. Cherokee Six One Tango. Good day. [Now change the transponder to 1200.]

DEPARTING AN AIRPORT UNDERLYING A CLASS B AIRSPACE

You're departing Spirit of St. Louis, Kansas City Downtown, Peachtree DeKalb (Atlanta), or any other field that lies under but is not in a Class B airspace. What differences, if any, does this situation pose as far as clearances, radio contacts, and the like are concerned?

Assume that you're leaving Spirit, which is west-southwest of St. Louis's Lambert Field, your route of flight is to the northeast, and your desired altitude will put you squarely in the Class B airspace. Also, assuming in this case that the clearance delivery service is *not* available (although in actuality it does exist at Spirit of St. Louis) the sequence of contacts would go like this:

You: Spirit Ground, Cherokee One Four Six One Tango at the terminal, ready to taxi, VFR northeast with Information Delta.

CD: *Cherokee One Four Six One Tango, taxi to Runway Two Six left.*

The runup is complete, you've taxied to the hold line, the transponder is on 1200, the switch in STANDBY:

You: Spirit Tower, Cherokee One Four Six One Tango, ready for takeoff, northeast departure.

Twr: *Cherokee Six One Tango, cleared for takeoff. Northeast departure approved.*

You: Roger, cleared for takeoff, Cherokee Six One Tango.

While taxiing onto the runway, change the transponder switch to ALT and get ready to roll. When you are safely off the ground, the tower should say:

Twr: *Cherokee Six One Tango, contact St. Louis Approach, one two six point seven.*

You: Roger. One two six point seven, Cherokee Six One Tango.

Now, dial in 126.7—if you didn't know the frequency in advance—and call Approach:

You: St. Louis Approach, Cherokee One Four Six One Tango.

Using Approach/Departure Control in the Controlled Airport Environment

App: *Cherokee Six One Tango, St. Louis Approach.*

You: Approach, Cherokee Six One Tango just off Spirit, heading of zero three zero at one thousand eight hundred, requesting seven thousand five hundred to Decatur and clearance through Bravo airspace.

App: *Cherokee Six One Tango, squawk zero two five six and ident. Remain outside the Bravo airspace until radar contact.*

You: Cherokee Six One Tango squawking zero two five six. Remaining clear.

You must contact Approach Control at the primary Class B airport for clearance into the airspace. (Remember why it's Approach and not Departure? If your recollection is faint, go back to Chap. 12.) If you do forget when the time comes and happen to call Departure instead of Approach, the controller won't subject you to public reprimands. Should you say "St. Louis Departure" and the controller acknowledges with "St. Louis Approach," go ahead with what you wanted to say and henceforth refer to the service as "Approach."

An important point here: The fact that Approach has given you a discrete squawk and asked you to ident does not constitute clearance into the airspace. Whatever you do, don't plunge merrily along. Level off to avoid busting the 2,000-foot floor; circle, do S-turns, or anything else that will keep you out of the sacred area until you hear something like this:

App: *Cherokee Six One Tango, radar contact. Cleared into the Bravo airspace. Turn right, heading zero four zero. Climb and maintain three thousand five hundred.*

You: Roger, understand cleared into Bravo. Right to zero four zero. Out of one thousand eight hundred for three thousand five hundred. Cherokee Six One Tango.

A word about altitude reporting: When working with Departure or Approach, always advise when you're leaving one altitude for a newly assigned altitude. The only time you don't have to worry about such reports is when you are advised to maintain VFR altitudes at your discretion. "At your discretion" means that you can climb or descend as you desire. In a controlled environment, however, you must report altitude changes and maintain the last one assigned. Unless specifically requested by the controller, you are not required to report reaching your new altitude, although some controllers prefer that you do. If for any reason you can't stay VFR at the assigned altitude, advise Approach or Departure and request a new altitude. Jumping around horizontally or vertically on your own in a controlled airspace such as a Class B or C is very, very *verboten*. Unless you hear "at your discretion," stick to the altitude and heading assigned.

To continue the illustration:

You: Approach, Cherokee Six One Tango level at three thousand five hundred.

App: *Cherokee Six One Tango, Roger. Traffic at twelve o'clock, three miles southeast bound at three thousand.*

You: Negative contact. Cherokee Six One Tango.

Perhaps a couple of minutes later:

App: *Cherokee Six One Tango, traffic no longer a factor. Climb and maintain seven thousand five hundred.*

You: Roger, out of three thousand five hundred for seven thousand five hundred. Cherokee Six One Tango.

When at the assigned altitude:

You: Approach, Cherokee Six One Tango level at seven thousand five hundred.

App: *Cherokee Six One Tango. Roger.*

When you are clear of the Bravo airspace, Approach will advise you accordingly:

App: *Cherokee Six One Tango, position two zero miles northeast of St. Louis, leaving Bravo airspace. Squawk one two zero zero. Radar service terminated. Frequency change approved. Resume own navigation.*

You: Roger. Cherokee Six One Tango. Thanks for your help. Good day.

DEPARTING A CLASS C AIRPORT

A Class C Departure offers two possible advantages for the VFR pilot over a Class B airspace. First, if you have the traffic ahead of you (or the traffic you are supposed to follow) in sight and can maintain visual separation, Departure might clear you to turn on course and climb to your cruising altitude. In other words, once that clearance is issued, your departure is much like departing a typical, nonradar, tower-controlled airport. There would be minimum vectoring, if any, and ATC would contact you only to give you advisories of other traffic that might be in your vicinity. The principle of visual separation for VFR aircraft would be in operation. All of this means, of course, that your flight would be more direct and the departure more rapid. This might not always be the standard procedure. Much depends on weather conditions, visibility, and the volume of traffic in the airspace. It is, however, a prerogative of ATC, when circumstances permit.

A second advantage for the VFR pilot who wants to get up and go is the size of the airspace. With only a 10-mile radius to the limits of the outer circle, compared to 15 to 30 miles for a Class B airspace you clear the regulated airspace much more rapidly and are free to fly on your own or contact Center, if you wish.

Clearance Delivery

Most Class C airspaces, but not all, have Clearance Delivery (the frequency is in the *Airport/Facility Directory*). If no Clearance Delivery exists, merely contact Ground Control and communicate your intentions. Ground provides the same service as Clearance Delivery and then clears you to taxi.

Whichever the case, the radio calls are identical. Listen to the ATIS, and then tune to the Clearance frequency. To illustrate the communication one more time, let's say you're leaving the Birmingham, Alabama, Class C airspace for Memphis:

You: Birmingham Clearance, Cherokee One Four Six One Tango at Hangar One with Echo. VFR Memphis via Victor 159, requesting six thousand five hundred.

CD: Cherokee One Four Six One Tango, Birmingham Clearance, Cleared to enter the Charlie airspace. Maintain three thousand. Departure frequency one two four point five, squawk two six three six.

You: Roger, cleared to enter the Charlie airspace, three thousand, one two four point five, and two six three six. Cherokee Six One Tango.

CD: Cherokee Six One Tango, readback correct. Contact Ground.

Next, the usual call to Ground and then to the tower is made.

Tower and Departure Control

You're at the hold line and ready to go:

You: Birmingham Tower, Cherokee One Four Six One Tango ready for takeoff.

Twr: Cherokee Six One Tango, cleared for takeoff. Fly runway heading.

In this transmission, the tower would probably instruct you to "contact Departure" when airborne. If not, be sure to request the frequency change after you are off the ground and have the aircraft under control. Don't go from one frequency to another in a controlled area without permission.

The call to Departure is much the same as in a Class B airspace:

You: Birmingham Departure, Cherokee One Four Six One Tango is with you, out of one thousand two hundred for three thousand.

Dep: Cherokee Six One Tango, turn right heading three one zero. Report reaching three thousand.

You: Right to three one zero and report three thousand. Cherokee Six One Tango.

A few minutes later:

You: Birmingham Departure, Cherokee Six One Tango level at three thousand.

Dep: Cherokee Six One Tango, Roger. Climb and maintain six thousand five hundred.

You: Roger, out of three thousand for six thousand five hundred. Cherokee Six One Tango.

When at altitude:

> **You:** Birmingham Departure, Cherokee Six One Tango level at six thousand five hundred.

> *Dep: Cherokee Six One Tango, Roger.*

When you pass the 10-mile outer circle, you can request termination of radar advisories, if you desire. Otherwise, radar service continues until you reach the 20-mile perimeter of the outer area, when you hear the final word from Departure:

> *Dep: Cherokee Six One Tango, leaving Charlie airspace, radar service terminated. Squawk one two zero zero. Change to advisory frequency approved.*

> **You:** Roger, Cherokee Six One Tango. Good day. [Be sure to change the transponder to 1200.]

Between these exchanges might come vectors, advisories, or even safety alerts, depending on conditions or traffic within the airspace. Or, as mentioned earlier, if the visibility is good and the traffic light, Departure Control has the prerogative to authorize a departing VFR aircraft to proceed directly on course and to continue its climb to the desired cruising altitude. That does not obviate the need, however, for the pilot to maintain radio contact with Departure until he is clear of at least the outer circle's 10-mile radius. In the outer area, which is not literally part of the Class C airspace, radar service can be declined by the pilot but not by ATC.

Departing a Satellite Airport within the Inner Circle

If you're departing a no-tower airport that is within the airspace, say, four miles from the primary airport, and you can't reach the primary airport's tower by radio on the ground, you can still take off, but you must follow the established traffic pattern for the satellite, and you must contact the tower as soon as possible after takeoff.

DEPARTURE FROM OTHER RADAR AIRPORTS

This time you're leaving from a Class D airport such as Springfield, Missouri. Like Clarksburg in Chap. 11, there are quite a few Class D airports around that also offer radar Approach and Departure Control.

To avoid repetition, let's assume that you have the ATIS at one of these airports, have been cleared by Ground to the hold area, and are ready to contact the tower:

> **You:** Springfield Tower, Cherokee One Four Six One Tango ready for takeoff, north departure.

> *Twr: Cherokee Six One Tango, cleared for takeoff. North departure approved.*

You're now off the ground:

> **You:** Springfield Tower, Cherokee Six One Tango requests frequency change to Departure.

Twr: *Cherokee Six One Tango, frequency change approved.*

So you thought the tower would automatically give you the Departure frequency? Wrong. You should have done your homework and determined the frequency before you got in the airplane. That's just one small element of preflight preparation—preparations that the amateur neglects and the professional doesn't forget. But you've screwed up a little, so:

You: Tower, Cherokee Six One Tango. What is the Departure frequency, north?

Twr: *Cherokee Six One Tango, contact Departure on one two four point niner five.*

You: One two four point niner five. Cherokee Six One Tango. Thank you.

You: Springfield Departure, Cherokee One Four Six One Tango is four north of Springfield, out of two thousand for six thousand five hundred, VFR Kansas City, squawking one two zero zero. Request traffic advisories.

Dep: *Cherokee Six One Tango, squawk zero two zero zero and ident.*

You: Cherokee Six One Tango squawking zero two zero zero.

Dep: *Cherokee Six One Tango, radar contact. Report reaching six thousand five hundred.*

From here on, Departure might vector you, alert you to other aircraft, or do whatever else is necessary to clear you from the area. When you reach your altitude, however, don't forget to communicate that fact:

You: Springfield Departure, Cherokee Six One Tango. Level at six thousand five hundred.

Dep: *Cherokee Six One Tango, Roger.*

Eventually, you'll hear this:

Dep: *Cherokee Six One Tango, radar service terminated. Squawk one two zero zero. Change to advisory frequency approved.*

You: Roger. Cherokee Six One Tango. Good day.

Now you're free to turn to any radio frequency you want—Flight Service, Flight Watch, unicom of any airports you'll be passing over, a tower or ATIS of a larger airport on your route, etc. The choice is yours.

Conclusion

That fairly well sums up the Departure Control procedures. As is hopefully apparent, there's nothing complicated about the process or what's required of the pilot. If you've done just a little homework ahead of time, if you know the frequencies, know what to say to what facility, and have a reasonably good idea of what you can expect to hear, getting out of even the busiest Class B is not difficult. Beyond those basics, once contact is made with each facility—Clearance, Ground, Tower, and Departure—it's really no more than a matter of listening, following instructions, and paying attention to what's going on outside the cockpit, none of which should be a challenge to any pilot.

But now that you're out of the controlled airport environment, the next matter of concern is how to get back into one. So with that in mind, let's turn to Approach Control and its role in the scheme of things.

APPROACH TO A CLASS B AIRPORT

The various radio procedures should be fairly well in your mind by now, and those required for entering a Class B are only minor variations of several that I have already illustrated. Regardless, let's start at the beginning just to be sure that the entire range of communications has been covered.

Before calling on Approach for radar assistance into any airport, you have to be prepared to supply certain information. To do that, keep in mind the so-called acronym suggested in Chap. 11:

I   = Aircraft *Identification*—type and full N-number

P   = *Position*—where you are geographically

A   = Present *Altitude*

I/D = *Intentions* and/or *destination*

S   = *Squawk*—what code you are currently squawking.

A   = *ATIS* information

The reason for "*Intentions* and/or *Destination*" is that the two might not be the same. For instance, you're approaching Dallas/Ft. Worth from the east enroute to Abilene, Texas. You merely want (or intend) to cross through the Bravo airspace, and the fact that Abilene is your ultimate destination is of little importance to Dallas Approach. It's not wrong to include "Abilene" in the radio call, but the main thrust of the call is your *intention* to transit the airspace, not your final destination. In reality, the I/D portion of the acronym can be stated one of three ways:

1. "Request clearance into and through the Bravo airspace to the west"
2. "Request clearance into and through the Bravo airspace to the west to Abilene"
3. "Request clearance into and through the Bravo airspace to Abilene."

Of course, if you intend to land anywhere in the airspace, as Ft. Worth, or in the Bravo's immediate area, that's your destination, and the name of the airport is what Approach needs to know. The I/D portion of the call would then be
"Landing Ft. Worth Meacham."

Keeping this sequence in mind, and rehearsing what you're going to say before getting on the mike, conveys an image of competence—which, in turn, increases the likelihood of receiving the assistance that you want. Controllers like to deal with pilots who sound as though they know what they're doing.

Using Approach/Departure Control in the Controlled Airport Environment

Let's use St. Louis as an example again. You're coming in from the west intending to land at Lambert Field, a Class B airport and are approaching the Foristell VOR, which is about 30 statute miles west of Lambert and about 10 miles from the outermost Class B ring. Now is the time to check the ATIS and call Approach:

You: St. Louis Approach, Cherokee One Four Six One Tango.

App: *Cherokee One Four Six One Tango, St. Louis Approach.*

You: Cherokee One Four Six One Tango, five miles west of Foristell VOR at seven thousand five hundred, landing Lambert. Squawking one two zero zero with Information Lima.

App: *Cherokee Six One Tango, squawk zero two three five and ident. Remain outside Bravo airspace until radar contact.*

You: Cherokee Six One Tango squawking zero two three five. Remaining clear.

Remember that just being given a squawk code does not authorize you to enter the airspace nor does the controller's statement that radar contact has been established. It's only when you hear "cleared to enter Bravo airspace" that you can penetrate that airspace. "Cleared" is the key word. (Sorry to beat this point, but the FAA is very emphatic about it.)

App: *Cherokee Six One Tango, radar contact. Cleared to enter Bravo airspace. Descend and maintain four thousand five hundred.*

You: Roger, understand cleared to enter the airspace. Out of seven thousand five hundred for four thousand five hundred, Cherokee Six One Tango.

A few minutes later:

You: Approach, Cherokee Six One Tango level at four thousand five hundred.

App: *Cherokee Six One Tango. Roger. Turn left, heading zero eight five.*

You: Left to zero eight five, Cherokee Six One Tango.

From this point on, Approach might give various instructions relative to new headings, altitude changes, advisories of other traffic in or near your line of flight, and the like. Each instruction or advisory should be promptly acknowledged—and not just with "Roger." Remember to repeat tersely whatever the instruction is or to advise Approach if you see or don't see the traffic that is in your vicinity.

Eventually, after you've been vectored and granted permission to descend further, Approach turns you over to the tower:

App: *Cherokee Six One Tango, contact St. Louis Tower on one one eight point five.*

You: Roger. One one eight point five. Cherokee Six One Tango.

Now what do you tell the tower? Almost nothing. The controller knows you're coming because Approach has so advised him. Thus there's no need for a time-consuming position/altitude/squawk report. Just this:

> **You:** St. Louis Tower, Cherokee One Four Six One Tango is with you, level at two thousand.
>
> **Twr:** *Cherokee Six One Tango. Enter left base for Runway Three Zero Left.*
>
> **You:** Roger, left base for Three Zero Left. Cherokee Six One Tango.

The instructions and landing continue in the normal pattern. To repeat some principles:

- Never leave an assigned altitude in a B airspace without permission.
- Always report when leaving one altitude for another that has been newly assigned. Report reaching the new altitude if you are instructed to do so (otherwise this is optional).
- "At pilot's discretion" means to climb or descend when you wish.
- When told to "ident," you don't need to acknowledge with "Cherokee Six One Tango identing." Just push the button. Do not push the button, though, unless you are instructed to ident.
- Don't change from an assigned squawk back to 1200 until you are told.
- If, for any reason, you can't remain at the assigned altitude because of weather, request permission to climb or descend, but always remain VFR!

> **You:** Approach, (Remember that VFR traffic *must* stay clear of clouds in Class B airspaces.) Cherokee Six One Tango requests lower account ceilings.
>
> **App:** *Cherokee Six One Tango, descend and maintain two thousand five hundred.*
>
> **You:** Leaving three thousand five hundred for two thousand five hundred, Cherokee Six One Tango.

One other point is appropriate to emphasize in this discussion: When Approach, Departure, or tower gives you an instruction, obey it immediately. For example, Approach tells you to turn to a heading of 120 degrees. As soon as you hear the heading, begin your turn. Don't keep flying on your present course while you pick up the mike (perhaps having to wait until the air is clear) to acknowledge the instruction. Do it now. Then, while in the turn, call Approach and confirm the 120-degree heading. Similarly, if you're told to climb or descend, obey immediately and confirm that action as soon as you can. The controller probably has a good reason for varying your line of flight, so don't hesitate to do what you're told.

APPROACHING AN AIRPORT THAT UNDERLIES A CLASS B AIRSPACE

Instead of landing at Lambert Field, you want to go into Spirit of St. Louis—which underlies Lambert's airspace. This time, however, you're coming in from the northeast and would like vectors through the airspace rather than detouring around it. Despite the fact that you're landing at a Class D airport, the radio communications are much the same.

Using Approach/Departure Control in the Controlled Airport Environment

To set the scene, you're cruising at 6,500 feet and tracking the St. Louis VORTAC, which is about 10 statute miles (sm) northwest of Lambert. Well outside the Class B airspace, make your first call:

You: St. Louis Approach, Cherokee One Four Six One Tango.

App: *Cherokee One Four Six One Tango, St. Louis Approach.*

You: Approach, Cherokee One Four Six One Tango is over Bunker Hill on the St. Louis zero five niner radial, level at six thousand five hundred, landing Spirit. Squawking one two zero with Information Papa.

App: *Cherokee Six One Tango, squawk zero two five two and ident. Remain outside Bravo's airspace until radar contact.*

You: Cherokee Six One Tango, zero two five two, remaining clear.

App: *Cherokee Six One Tango, radar contact three zero miles east of the St. Louis VOR. Cleared to enter Bravo airspace. Descend and maintain four thousand five hundred. Turn left, heading one niner zero.*

You: Roger, understand cleared to enter Bravo airspace. Out of six thousand five hundred for four thousand five hundred, left to one niner zero. Cherokee Six One Tango.

A little later:

App: *Cherokee Six One Tango, turn right, heading two seven Zero.*

You: Right to two seven zero, Cherokee Six One Tango.

App: *Cherokee Six One Tango, descend and maintain two thousand five hundred.*

You: Out of four thousand five hundred for two thousand five hundred. Cherokee Six One Tango.

As you near the Spirit Airport, you'll be below the 3,000-foot airspace floor. About this time, Approach will call you:

App: *Cherokee Six One Tango, position five miles east of Spirit. Contact Spirit Tower, One Two Four point Seven Five.*

You: One Two Four point Seven Five. Thank you for your help. Cherokee Six One Tango.

Now contact the Spirit Tower for pattern and landing instructions: "Spirit Tower, Cherokee One Four Six One Tango is with you 5 miles east, level at two thousand five hundred."

Using the same situation of an airport underlying a Class B, let's suppose that you are not familiar with the airport and there is no VOR or nondirectional beacon (NDB) on the field for guidance. In other words, you need assistance.

You're again coming into St. Louis from the northeast. After radar contact has been established and you are cleared into the Class B airspace, Approach calls you: "Cherokee Six One Tango, cleared on course. Descend and maintain two thousand five hundred."

Being uncertain of the location of the airport, you request further help:

You: Roger. Cherokee Six One Tango is out of three thousand five hundred for two thousand five hundred. Request vectors to Spirit.

App: *Cherokee Six One Tango, Roger. Turn right heading two seven zero to Spirit.*

You: Right to two seven zero. Cherokee Six One Tango.

App: *Cherokee Six One Tango, Spirit is at twelve o'clock, eighteen miles. Report when you have it in sight.*

You: Will do, Cherokee Six One Tango.

In a few minutes:

You: Approach, Cherokee Six One Tango has Spirit in sight.

Approach then turns you over to Spirit Tower as before.

Another situation isn't unusual when the Class B is busy and your destination is an underlying airport: permission to enter the airspace is not granted. In that case, getting into a field such as Spirit from the northeast means skirting the area. It adds time and mileage, but you have no alternative:

You: St. Louis Approach, Cherokee One Four Six One Tango.

App: *Cherokee One Four Six One Tango, St. Louis Approach.*

You: Approach, Cherokee Six One Tango is over Bunker Hill at six thousand five hundred, squawking one two zero zero with Information Papa. Request clearance into the Bravo airspace for landing Spirit.

App: *Cherokee Six One Tango, unable. Change to Spirit frequency one two four point seven five and remain clear of the Bravo airspace.*

You: Roger, Approach. Cherokee Six One Tango remaining clear.

With those instructions, your only alternative is to detour around the airspace and stay beneath all floors until the Spirit Airport is in sight. At that point, make the routine call to Spirit's Tower.

TRANSITING A CLASS B AIRSPACE

This time, you don't want to land in or under the airspace but instead pass through it to a more distant destination. What you say and what you do are almost identical to landing procedures. Assume that you're east of St. Louis and are heading for Jefferson City, Missouri. The straight line from your present position would put you right through Bravo airspace. You could go around the whole thing, but, again, that would only add time and fuel costs—so why not take the economical way out? That being the more intelligent avenue, you plan what you're going to say and call Approach:

You: St. Louis Approach, Cherokee One Four Six One Tango.

App: Cherokee One Four Six One Tango, St. Louis Approach.

You: Approach, Cherokee One Four Six One Tango approaching Troy VOR on zero seven six radial, level at six thousand five hundred to Jeff City. Squawking one two zero zero. Request vectors through Bravo airspace.

App: Cherokee Six One Tango, squawk zero two six four and ident. Remain outside Bravo airspace until radar contact.

You: Cherokee Six One Tango squawking zero two six four, and remaining clear.

App: Cherokee Six One Tango, radar contact. Cleared to enter Bravo airspace. Maintain six thousand five hundred. Turn right, heading two seven zero.

A pause here to explain something: You reported your position as "approaching Troy VOR on zero seven six radial." Why the "076 radial?" Approach knows, of course, where the Troy VOR is located, so it should be easy to spot your aircraft once you've squawked the discrete code. Disregarding that, however, give the radial *from* the VOR in a position report. That makes identification of your aircraft much easier for the controller. "Approaching on the zero seven six radial" alerts him or her to look for you on the east side of the screen. In this case, the 076 radial also happens to be the centerline of Victor 12, a VOR airway. So you could alternatively report "five miles east of Troy on Victor Twelve."

If, at any time, you are asked by Center or Approach to advise what *radial* you are flying, report it in terms of the *outbound bearing* from the station. If asked your *heading*, report the actual reading on your compass or directional gyro.

Now back to the example:

You: Roger, understand cleared to enter Bravo. Maintain six thousand five hundred, right to two seven zero. Cherokee Six One Tango.

A few minutes later:

App: Cherokee Six One Tango, turn left, heading two four five.

You: Left to two four five. Cherokee Six One Tango.

App: Cherokee Six One Tango, traffic at eleven o'clock, two miles northeast bound at five thousand five hundred.

You: Roger, negative contact. Cherokee Six One Tango.

With your eyes scanning the skies at about the eleven o'clock position (don't forget to compensate for any crab angle), you spot the traffic approaching your position, but below you:

You: Approach, Cherokee Six One Tango has the traffic.

App: *Cherokee Six One Tango, Roger.*

Eventually, after whatever vectoring Approach deems necessary to provide the proper aircraft separation, you are clear of Bravo on the west side:

App: *Cherokee Six One Tango, position two zero miles southwest of St. Louis VOR. Departing Bravo airspace. Squawk one two zero zero. Radar service terminated. Frequency change approved.*

You: Roger. One two zero zero—and thank you for your help. Cherokee Six One Tango.

You're on your own again, with Jeff City off in the distance.

APPROACH CONTROL AND THE CLASS C AIRSPACE

The mandatory Approach Control radar service at a Class C airspace requires no new or additional radio phraseology. The differences between the Class B and Class C airspaces are confined to procedures or regulations meaning:

Class B airspaces: Clearance into a Class B airspace is required; clearance can be denied.

Class C airspaces: Radio contact with Approach must be established before entering the Class C airspace, but a literal "clearance" into the airspace is not required; use of radar service is mandatory.

If you keep IPAI/DS in mind and have the basic Class B call pattern mastered (plus, of course, two-way radio communications capability and Mode C), you have all you need to operate in a Class C airspace.

To follow through with an example: You're approaching the Tucson, Arizona, Class C from the northwest on Victor Airway 308 for landing at Tucson International. About 20 miles out, you contact Approach. Remember that you can enter a Class C outer area without radio contact but not the 10-mile radius outer circle.

You: Tucson Approach, Cherokee One Four Six One Tango.

App: *Cherokee One Four Six One Tango, Tucson Approach.*

You: Approach, Cherokee Six One Tango is twenty northwest on Victor 303, level at seven thousand five hundred for landing International, squawking one two zero zero with Information Echo.

App: *Cherokee Six One Tango, squawk four five four three and ident.*

You: Four five four three. Cherokee Six One Tango.

App: *Cherokee Six One Tango, radar contact. Fly present heading and descend and maintain six thousand.*

You: Roger, Approach, Six One Tango leaving seven thousand five hundred for six thousand.

Along with altitude changes and possible, or probable, vectors, you finally hear:

App: *Cherokee One Four Six One Tango, contact International Tower one one eight point three.*

You: Roger, one one eight point three, Six One Tango.

You: International Tower, Cherokee One Four Six One Tango is with you, level at four thousand five hundred. (Tucson's elevation is 2,641 feet msl, with a light aircraft pattern altitude of 3,400 feet msl.)

Twr: *Roger, Cherokee One Four Six One Tango, enter left base for Runway Two Niner.*

You: Left for Two Niner. Six One Tango.

And so it goes until you're down and parked.

Suppose, however, in response to your introductory call, that Approach came back with this:

App: *Cherokee One Four Six One Tango, stand by.*

What do you do now? Nothing. Stay right on course into the airspace. You have established the required communications, which is all that's necessary. Unless the controller: (1) doesn't answer you at all; (2) replies with "Aircraft calling Tucson Approach, stand by"; or (3) specifically calls you by your aircraft N-number and tells you to "remain clear (or outside) of the Class C airspace," you can legally enter the area. Therein lies a major Class B versus Class C difference.

TRANSITING A CLASS C

If you're on a cross-country at normal VFR cruising altitudes in the 6,500- to 8,500-foot range, you won't have much trouble staying above most Class C airspaces. Their typical 4,000-foot-plus agl ceiling makes overflying relatively easy—unless, of course, weather forces a lower altitude. In such instances, transiting a Class C that lies along your route might be necessary to avoid a flight-prolonging detour.

Fortunately, the radio contact is almost identical to that when transiting a Class B. To illustrate, assume that you're enroute from Memphis to the Peachtree Dekalb Airport near Atlanta, and the cloud cover has forced you to cruise at 3,500 feet msl. Nearing the Birmingham, Alabama, Class C airspace, you make the initial call:

You: Birmingham Approach, Cherokee One Four Six One Tango.

App: *Cherokee One Four Six One Tango, Birmingham Approach.*

You: Approach, Cherokee One Four Six One Tango is twenty northwest on Victor One Five Niner, level at three thousand five hundred, enroute Peachtree Dekalb, squawking one two zero zero. Request advisories through the Charlie airspace.

> **App:** *Roger, Cherokee Six One Tango. Squawk four two five three and ident.*

> **You:** Cherokee Six One Tango squawking four two five three.

> **App:** *Cherokee Six One Tango, radar contact. Birmingham altimeter is Two Niner Four Seven.*

> **You:** Roger, Two Niner Four Seven. Cherokee Six One Tango.

What happens for the next few minutes depends on the volume of traffic in the airspace and the advisories or safety alerts ATC issues you. But what would you do if ATC, for good reasons, asks you to leave your present altitude and climb to one that would put you in or close to the base of the overcast ceiling? You'd then be less than 500 feet below the cloud layer and in violation of VFR regulations.

The simple answer, and the point for reemphasis is: Don't blindly follow orders that would result in a violation. Advise ATC of the situation. The FARs make it very clear that you're in command of the aircraft. Yes, directions should be obeyed, but ATC might not be aware that an altitude change would make you a rule-breaker. Consequently, if any instruction would cause you to violate a regulation, regardless of the type of airspace, make that fact known to the controlling agency.

For altitude separation purposes between a VFR and an IFR aircraft, you might be told to climb or descend to a non-VFR cruising altitude, such as 3,000 or 4,000 feet. When the altitude separation is no longer needed, and especially when leaving the regulated airspace, ATC will advise you to "Resume appropriate VFR altitudes." That's your directive to climb or descend to the odd- or even plus-500-feet VFR cruising altitudes dictated by FAR 91.159.

Returning to the Birmingham exchanges, when you depart the outer area on your way to Atlanta, you'll eventually hear:

> **App:** *Cherokee Six One Tango, leaving the Charlie airspace, radar service terminated. Squawk, one two zero zero. Change to advisory frequencies approved.*

> **You:** Roger, Cherokee Six One Tango, good day. [Now change the transponder to 1200.]

APPROACH CONTROL AT CLASS D TOWER-ONLY AIRPORTS

At those Class D airports that have only a tower but no Class B or Class C airspace, such as the Clarksburg example described in Chap. 11, you have the option to use or not use Approach Control. One situation in which Approach at such airports can be helpful (other than providing advisories, of course) is when you're entering unfamiliar territory and would like directions (vectors) to the airport area. Using Clarksburg again, the following simulates the radio communications for an approach and landing at the city's Benedum Airport:

> **You:** Clarksburg Approach, Cherokee One Four Six One Tango.

> **App:** *Cherokee One Four Six One Tango, Clarksburg Approach.*

You: Approach, Cherokee One Four Six One Tango is over New Martinsville level at five thousand five hundred, landing Benedum. Squawking one two zero zero with Foxtrot.

App: *Cherokee Six One Tango, squawk four five five three and ident.*

You: Four five five three, Six One Tango. Request.

App: *Cherokee Six One Tango go ahead.*

You: Am unfamiliar with the area, Approach, and request vectors to the airport.

App: *Roger, Six One Tango. Radar contact 30 miles northwest. Fly heading one three zero, vectors later to Benedum.*

You: Roger, one three zero. Six One Tango.

Later:

App: *Cherokee One Four Six One Tango, turn left, heading one two zero. Descend pilot's discretion and maintain three thousand.*

You: Left to one two zero, and leaving five thousand and five for three thousand. Cherokee Six One Tango.

You: Approach, Cherokee Six One Tango level at three thousand.

App: *Roger, Six One Tango. Benedum is at your one o'clock position, seven miles. Report the airport in sight.*

You: Approach, Six One Tango has the airport.

App: *Roger, Six One Tango. Contact Benedum Tower on one two six point seven.*

You: One two six point seven. Thank you. Cherokee Six One Tango.

You: Benedum Tower, Cherokee One Four Six One Tango is with you, level at three thousand.

Twr: *Cherokee One Four Six One Tango, enter right base for landing Runway Two One.*

You: Right for Two One, Cherokee Six One Tango.

A few minutes later:

Twr: *Cherokee Six One Tango, cleared to land Two One.*

You: Cleared to land. Six one Tango.

CONCLUSION

The only challenge that Approach and Departure Control pose is the challenge of the unknown. When you're insecure or uncertain about what to do or say, you'll probably take evasive action to avoid an upcoming Class B (Bravo) or Class C (Charlie) airspace. As a result, you won't go places you'd like to go; you'll add needless miles and hours to an otherwise straight-line cross-country flight; you'll land at out-of-the-way airports and hope that some form of transportation will be available to get you to your ground destination. Or you'll bounce around at 2500 feet to stay under a Class B airspace when you could have a smooth ride at 6500 feet in the Bravo airspace. And so on...all because you're leery about contacting Approach or Departure Control. To do so, you may feel, would be beyond your scope of knowledge or certainly your scope of confidence.

Such need not be the case. To repeat: Using Approach or Departure Control simply means knowing what you're going to say, why you should say it, and being aware of the general thrust of what you're likely to be told. Once you have the IPAI/DS' format mastered, the rest is just a matter of listening, acknowledging, and obeying. ATC tells you what to do; obedience within the limits of safety and flight rules and keeping your eyes open are then your responsibility.

So use the services that are available to you as a VFR pilot. Know what you're going to say; say it confidently, tersely, and distinctly; say it with the ring of a professional. As so many pilots have learned, when their workload permits, you'll get all the help you need from those folks on the ground.

14
Working with Air Route Traffic Control Centers on VFR Flights

MOST OF THE PREVIOUS CHAPTERS HAVE KEPT YOU IN THE GENeral vicinity of the airport—Tower, Approach, Departure, and the like. Now is the time to stray beyond the local aerodrome and consider the pilot's good cross-country friend: The Air Route Traffic Control Center (ARTCC, or "Center," for short).

CENTER'S ROLE IN THE SCHEME OF THINGS

Its name describes its purpose and role: An ARTCC is indeed the *center* for control of all enroute IFR and, when its workload permits, advisories to VFR traffic, within its assigned geographical area. An ARTCC thus ensures the vertical and horizontal separation of IFR aircraft and alerts participating VFR aircraft to potential traffic hazards. In so doing, *control* is the important word. That's Center's job—which, as a rule, means that functions not directly related to the control of traffic, such as filing or amending a flight plan, requesting enroute weather, or asking for weather forecasts, should be directed to the nearest Flight Service Station.

This is not to say that a Center's responsibility is limited solely to enroute traffic control. Although it is not a forecasting agency, it can alert pilots to possible heavy precipitation that might lie just ahead; in a distress or urgency situation or when a pilot is lost and needs guidance to the nearest field, it can be a life-saver. And, of course, it assumes the Approach Control function at airports when the local Approach facility is closed or at other fields within its radar range that have no such facility at all.

Basically, though, Center is the bridge between Departure Control at the flight's airport of origin and Approach Control at point of landing. Once the aircraft is clear of the Class B or Class C airspace, Departure drops out of the picture and Center steps in. For the duration of the flight, one or more Centers (depending on the distance traveled) become the enroute controllers, assigning altitudes and vectors to IFR aircraft, advising participating VFR pilots of potential traffic, and generally monitoring the safety of the airways. As the flight nears its destination, Center either terminates the radar service (as it might for VFR aircraft) or hands the pilot off to Approach for sequencing and separation prior to the actual tower-controlled landing.

So, once again, the basic flow from taxi-out to landing at a Class B or C airport is as follows:

1. Clearance Delivery
2. Ground Control
3. Tower
4. Departure Control
5. Center
6. Approach Control
7. Tower
8. Ground Control

THE CENTERS AROUND THE COUNTRY

Twenty Air Route Traffic Control Centers are responsible for the enroute traffic in the contiguous 48 states, plus one each in Alaska, Hawaii, Puerto Rico, and Guam. By location, those in the lower 48 are:

Albuquerque	Houston	Minneapolis
Atlanta	Indianapolis	New York
Boston	Jacksonville	Oakland
Chicago	Kansas City	Salt Lake City
Cleveland	Los Angeles	Seattle
Denver	Memphis	Washington, DC
Fort Worth	Miami	

Geographically, the areas the various Centers control are illustrated in Fig. 14-1.

Fair enough, but what about the geography between almost any two of these locations? Take Washington, DC, and Jacksonville, for example. Referring to Fig. 14-1 again,

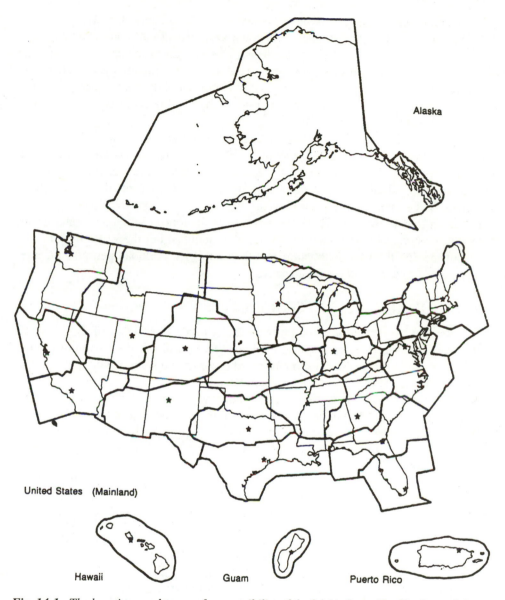

Fig. 14-1. *The locations and areas of responsibility of the 24 Air Route Traffic Control Centers.*

you'll see that the Washington Center (actually located at Leesburg, Virginia) controls traffic from just north of Washington through most of Virginia, down to the border between North and South Carolina. At that point, the Jacksonville Center takes over. But the two Centers are about 650 miles apart, so with that spread, control can be maintained only through a network of remoted communications that tie the more distant radar sites back to the physical location of each Center by microwave links or land lines.

If you were making this trip, would you talk to the same controller, even while still within Washington's geographical area? Most unlikely, because a Center is divided into sectors, with one or more controllers handling flights in a given sector. As you near the limits of one sector, the controller with whom you've been talking tells you to change to another frequency. This puts you in contact with a controller in the next adjacent sector— and so on through Washington's control area—each with a different frequency remoted back to the Leesburg ARTCC.

When you leave Washington's area, south of Wilmington, North Carolina, Washington gives you the first remoted Jacksonville ARTCC frequency to contact, which is at Myrtle Beach. In time, Jacksonville tells you to change frequencies again to the Savannah remoted site, and eventually, the Center turns you over to Jacksonville Approach for landing within the Jacksonville Class C.

Thus, throughout the country, you can always be in contact with one Center or another via this network of communication and radar facilities. For the VFR pilot, it's just a matter of making the first radio call. From then on, if its workload permits, a Center will help you from almost departure to arrival.

THE ENROUTE LOW ALTITUDE CHART

Although I could have mentioned the Enroute Low Altitude Chart (Fig. 14-2), or ELAC, at any time, I've reserved reference to it until now because of this discussion of Center, cross-country excursions, and ways to make flights easier and safer. If you're not familiar with the chart, stay with me for a brief summary.

Twenty-eight charts cover the country, each valid for a period of about two months. These are not 28 individual charts but rather two in one, as Fig. 14-3 indicates. L-21 extends from western Missouri east to portions of Kentucky. L-22 overlaps L-21 in Kentucky and continues east over parts of West Virginia and Virginia to the Atlantic coast.

L-22
PANELS
EFGH
1"=12 NM

L-21
PANELS
ABCD
1"=12 NM

UNITED STATES GOVERNMENT
FLIGHT INFORMATION PUBLICATION

IFR ENROUTE LOW ALTITUDE - U.S.

For use up to but not including 18,000' MSL
HORIZONTAL DATUM: NORTH AMERICAN DATUM OF 1983

EFFECTIVE 0901Z **17 JUL 1997**

TO 0901Z **11 SEP 1997**

Consult NOTAMs for latest Information

PUBLISHED IN ACCORDANCE WITH INTER-AGENCY AIR CARTOGRAPHIC COMMITTEE
SPECIFICATIONS AND AGREEMENTS APPROVED BY:
DEPARTMENT OF DEFENSE★FEDERAL AVIATION ADMINISTRATION★DEPARTMENT OF COMMERCE

Fig. 14-2. *Another aid for VFR navigation is the Enroute Low Altitude Chart (ELAC).*

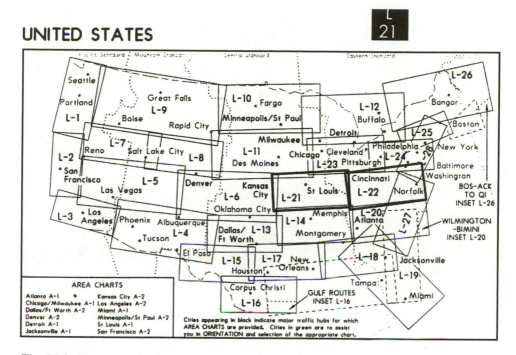

Fig. 14-3. *Twenty-eight charts cover the contiguous 48 states.*

It's called "Enroute Low Altitude" because it is for use up to but not including 18,000 feet msl. For the VFR pilot who wants to track VORs and use his radio effectively, it provides a wealth of information not found on the sectional. Figure 14-4 illustrates some of the typical data. Keep in mind that it is an *enroute* chart and thus omits some information found on the sectional, such as unicom frequencies. Nor does it depict topography, towns, roads, rivers, and similar landmarks.

Despite the absence of that information, the chart can almost totally replace the sectional during the enroute portions of a flight if you want to rely primarily on your navcom equipment for navigation. Indeed, it is for IFR flight and contains a lot of symbols that aren't pertinent to the VFR pilot. That fact doesn't reduce its usefulness, however.

I'm not suggesting that you discard the sectional, only that you use the two jointly. After all, you might lose radio contact and have to rely on the sectional to get you to the nearest airport. It's a good idea to use the charts in combination so that your position relative to identifiable ground references is never in question.

What do you find in the enroute that's not included in the sectional? A few examples are illustrated in Figs. 14-5 through 14-7. These represent some of the more meaningful data for the VFR pilot that make the chart a helpful reference and navigation resource. It's not an essential tool, and if you're going to travel strictly by dead reckoning, leave it home. Otherwise, plan your flight with both charts, and use both as you venture from here to there.

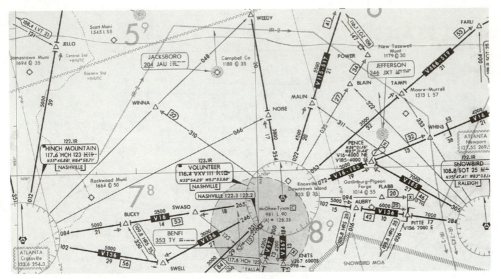

Fig. 14-4. *A sample of the ELAC data and how it is illustrated.*

THE ADVANTAGES OF CENTER FOR THE VFR PILOT

First, it should be made clear that enroute assistance from Center is neither mandatory nor always available for the VFR pilot. The primary purpose of Center is to facilitate the movement of IFR aircraft. It does not exist nor is it required to serve the VFR pilot. You can request enroute VFR advisories, but the controller has the right to refuse the request if his workload does not permit. It is very likely that in marginal VFR conditions or when there is a heavy concentration of traffic, your request for advisories will be rejected. At other times, you will often find Center's controllers not only helpful but eager to be of help.

A second observation: A Center controller is a professional who serves profession-als—meaning airline, corporate, and skilled instrument pilots. The vast majority of these airmen know (or should know) how to use their radios. The air-to-ground communica-tions are brief and to the point.

When a VFR pilot gets on the air and proceeds to stammer, hesitate, ramble, or gives the impression of incompetence, the likely response from Center to the request for en-route advisories will be "unable due to workload." That might not be the case at all, but the controller has concluded that he or she just doesn't have the time to fiddle around with an unprepared amateur. Nor can he be blamed. Remember the case of an airline pi-lot who reported his position in only five seconds while a private pilot took four minutes to convey the same basic information? Center has neither the time nor patience to ac-commodate that sort of amateurism.

None of this is intended to imply that the VFR pilot should avoid using Center or to intimidate the relatively inexperienced. The 100-hour pilot can sound just as professional as the 10,000-hour 747 captain. All it requires is practiced competence in the art of radio communications.

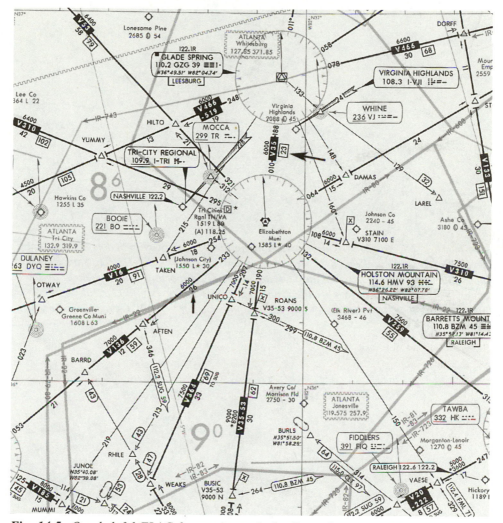

Fig. 14-5. *One helpful ELAC feature is nautical mileage between VORs, IFR reporting points, and airway intersections. The "23" on V35 (upper center) is the distance between the Glade Spring and Holston Mountain VORs. Southwest of the Holston Mountain VOR on V136, the "26" represents the mileage between the VOR and the AFTEN reporting point (see arrows).*

The advantages of using Center on a VFR cross-country, however, are enough to justify the development of professionalism. For instance:

- The controller with whom you are in contact is, in effect, another pair of eyes—groundbound though they may be. He or she has your airplane on his scope, identified by the discrete transponder code assigned you, and can thus alert you to traffic in your general vicinity. He can issue you warnings of military flight activity, such as B-52s on low-level training flights. If you want to change

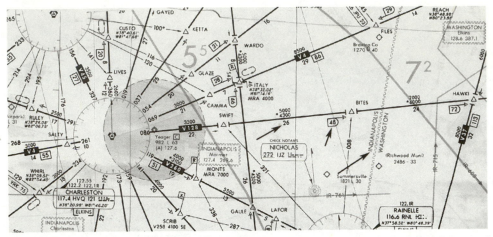

Fig. 14-6. *The "26" (slightly right of center) along the V128 airway indicates the nautical mileage between the SWIFT and BITES reporting points; the "48" is the DME (distance measuring equipment) mileage between the VOR and the BITES reporting point; the other arrow indicates the design used to identify the separation of Air Route Traffic Control Centers—in this case, where control passes from the Indianapolis Center to the Washington Center, or vice versa.*

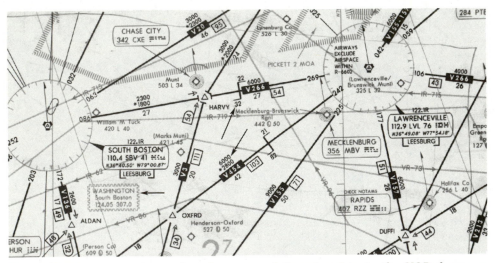

Fig. 14-7. *One arrow (approximately center of figure) points to the VOR frequency changeover point: 32 nautical miles to the South Boston VOR and 22 nautical miles to the Lawrenceville VOR. The fact that Washington is the responsible Air Route Traffic Control Center, through its South Boston RCO, is indicated by the box at the lower left. The minimum enroute altitude (MEA) for radio reception in this area on V454 is 6,000 feet and the minimum obstruction clearance altitude (MOCA) is 2,500 feet (see arrow, lower middle).*

altitudes, he or she can advise you of potential traffic at the new altitude you have chosen.

* If you lose your radio and squawk the 7600 radio failure code, the Center controller will spot the failure, can follow your course, and will notify subsequent controllers of the situation.

* If you encounter a serious emergency, you have someone on the frequency to whom you can talk now—no tuning to 121.5; no need to divert your attention from handling the emergency while you recode the transponder to 7700. The controller knows where you are and can lead you to the nearest airport, keep track of your position, and alert the sources closest to your position of your predicament.

In every respect, Center is an added insurance policy to your flight plan. As with all of the other radio aids, the service is there to be used. I must stress again, however, that in normal operations Center is not required to lend assistance to a VFR pilot. It's the one case where requested advisories can be rejected. The tower has no such freedom; Approach and Departure have no such freedom when you are landing or departing a Class B or C airport; nor does Ground Control. Only Center does. It must assist the IFR pilot but not those flying VFR.

As a VFR pilot, you can legally fly the length and breadth of the United States and never once contact any Center. But the proverbial question: Why not use the service, since it's available? At the very least, monitor the remoted frequencies as you pass from one area to another. Merely eavesdropping might alert you to a potential hazard somewhere in your line of flight. It's just as easy, however, to go whole-hog—to go first class—if you know what you're doing.

GOING FROM DEPARTURE TO CENTER

One of three VFR situations relative to Center can occur when Departure Control has been vectoring you out of any Class B, C, or D airspace. In the first situation, which is rather rare, Departure asks, as you're leaving the area, if you would like a "handoff" to Center for enroute traffic advisories. In other words, Departure is saying, "Would you like us to contact Center (by phone) to see if it can advise you of other traffic that might be in or near your line of flight as you proceed on your way?" The second situation is when you initiate the request. Departure then calls Center, and Center agrees to provide the service. In the third instance, you again request the handoff, but Center, for whatever reason, denies it. How Departure actually phrases its questions or statements will vary somewhat, but the dialogue generally follows these lines.

Situation 1

You are leaving the Kansas City Class B VFR for Denver, and about 30 miles out, Departure comes on the air:

> *Dep:* *Cherokee One Four Six One Tango, you are leaving Departure's radar area. Do you want enroute advisories?*

> **You:** Affirmative, Departure, if Center can handle. Six One Tango.

> *Dep:* *Roger, Six One Tango. Stand by.* [Pause, while Departure contacts by telephone the Center sector controller responsible for your geographic area.]

> *Dep:* *Cherokee Six One Tango, contact Kansas City Center, on one two zero point five. Good day*

> **You:** Roger, one two zero point five. Thank you. Cherokee Six One Tango.

Departure's last communiqué tells you that Center has accepted the handoff and will provide the traffic advisories you have requested. Also implied, but not stated, is the fact that once Center accepted the handoff, Departure gave the sector controller your aircraft type, N-number, present position, altitude, first point of landing, and squawk. In other words, Center knows a lot about you before you make the first contact—which is why that contact is merely "with you," plus your present altitude—or, if climbing, the desired cruising altitude. So, all you do now is leave the discrete transponder code where it is, change to the new frequency, and make the call:

> **You:** Kansas City Center, Cherokee One Four Six One Tango is with you, level at six thousand five hundred.

> *Ctr:* *Roger, Cherokee One Four, Six One Tango. Squawk two two four five and ident.*

> **You:** Two two four five. Cherokee Six One Tango.

> *Ctr:* *Cherokee One Four Six One Tango. Topeka altimeter three zero one zero. Current traffic C-130s in Topeka area practicing instrument approaches Forbes Field.*

> **You:** Roger, Center, will be looking. Cherokee Six One Tango.

What else follows depends on new traffic and advisories Center might want to communicate to you. Regarding the squawk code, a Center usually, but not always, assigns you a new computer-generated code. Until it does, though, leave the one that Departure gave you right where it is.

Situation 2

Here, Departure terminates radar coverage and makes no reference to Center, but you do want enroute traffic advisories:

> *Dep:* *Cherokee One Four Six One Tango, radar service terminated. Squawk one two zero zero. Frequency change approved. Good day.*

> **You:** Departure, Cherokee One Four Six One Tango requests handoff to Center, if possible.

> *Dep:* *Stand by, Six One Tango. We'll check...* [pause]...

> *Dep:* *Cherokee Six One Tango, contact Kansas City Center on one two zero point five.*

You: Roger, one two zero point five. Thank you. Cherokee Six One Tango.

That's all there is to it. Merely change the frequency and contact Center, just as in the previous example.

Situation 3

In this instance, you want enroute advisories, Departure requests it, but Center is unable to provide it:

Dep: *Cherokee One Four Six One Tango, radar service terminated, Squawk one two zero zero. Frequency change approved. Good day.*

You: Departure, Cherokee Six One Tango requests handoff to Center.

Dep: *Cherokee Six One Tango, stand by. We'll check…[pause]…*

Dep: *Cherokee Six One Tango, Center unable at this time. Radar service terminated, Squawk one two zero zero. Frequency change approved.*

You: Roger, squawking VFR. Thanks for your help. Cherokee Six One Tango.

With this, you're on your own as you head west. You can change altitudes at any time, as long as they conform to the VFR altitude regulations; you can alter your route of flight; you can change your radio to any frequency you wish; and you must remain VFR at all times. In other words, you have the freedom to do just about as you wish—as long as you abide by the VFR FARS.

Any time that you hear the word "terminated," you can assume that all radar service has ceased and that no handoff to the next agency has taken place. Now if you want Center or Approach, you have to give the full IPAI/DS. The next facility hasn't heard about you before.

Once in a while, a controller might misuse "terminated" as it's supposed to be applied per the book. You hear something like this from Departure: "Cherokee Six One Tango, radar service terminated. Contact Atlanta Center on one two four point seven." Has radar service literally been terminated, or have you been handed off to Center? The first part says terminated; the second part raises a question. Uncertain of what has really happened, you play the pro and call Center:

You: Atlanta Center, Cherokee One Four Six One Tango.

Ctr: *Cherokee One Four Six One Tango, Atlanta Center. Go ahead.*

The phraseology indicates that this is the initial contact with the controller, so you go ahead with IPAI/DS.

But suppose you hear this in response to that call:

Ctr: *Cherokee One Four Six One Tango, Atlanta Center. Squawk one two zero five and ident. Verify, altitude.*

You: Cherokee One Four Six One Tango, squawking one two zero five. Level at seven thousand five hundred.

This time the phraseology tells you that you have been handed off. In either case, once contact with Center has been established, the controller takes care of much of the conversation from that point on. Your basic job is to listen, respond, and acknowledge whatever information or instructions are conveyed.

Four Points about Handoffs—in Review

- The expression "with you" is used any time you are being handed off or automatically transferred from one agency to another: Departure to Center; Center to Approach; Approach to Tower.
- Even though it is a handoff, be sure to advise the receiving agency of your present altitude. If you fail to include it in your initial contact, you might be asked—which only adds to air clutter.
- Note that no position report is necessary. Center, or whatever the controlling agency is, knows your position as well as your destination and squawk.
- Don't change your transponder from one squawk to another until you are so advised. Let's say that Departure has given you 1201 and you have been handed off to Center. Keep 1201 in the box unless Center gives you a new code.

INITIATING THE CONTACT WITH CENTER

Either Departure hasn't handed you off or you're leaving an airport that doesn't have a Departure Control. Regardless of the situation, you want to establish the first contact with a Center for enroute advisories. You've determined the probable Center frequency through proper preflight preparation, so, again, it's the simple IPAI/DS:

You: Seattle Center, Cherokee One Four Six Tango.

Ctr: *Cherokee One Four Six One Tango, Seattle Center.*

You: Center, Cherokee Six One Tango is about 15 south of Astoria on Victor Two Seven, level at five thousand five hundred, VFR to Newport squawking One Two Zero Zero. Request traffic advisories, workload permitting.

Ctr: *Cherokee One Four Six One Tango, squawk four one four five and ident.*

You: Four one four five. Cherokee Six One Tango.

Ctr: *Cherokee Six One Tango, radar contact eighteen south of Astoria. Maintain VFR at all times. Astoria altimeter two niner seven five. Seattle Center.*

You: Roger, two niner seven five. And will maintain VFR. Cherokee Six One Tango.

That's all there is to it. Center will take it from there and keep you advised of what's going on around you.

About that comment, "workload permitting." It's obviously not required, any more than "good day" or "thank you," but it does put your request in the form of a request, not a command. In its way, it indicates appreciation of the fact that the controller might be busy and not able to provide the advisories. That little added phrase frequently is enough to get the help that otherwise might have been rejected. Call it courtesy and mutual understanding in the air.

Also, be sure that you have established contact with the Center before giving the IPAI/DS, as the initial transmission above indicates. The controller might be busy on another frequency and won't hear your call, so don't waste air time until you know you've got a listener (he or she might be monitoring up to six different frequencies at the same time).

Now that Center has you on radar and is tracking your progress, an admonition is in order: Even with Mode C, don't change altitudes without advising Center. The controller has you pegged at 5,500 feet, and if you don't have Mode C, there's no way he or she can determine your altitude without verification from you. Even with Mode C, an unapproved change to a different flight level might affect the flow of traffic and place you and others in jeopardy. You're not flying under instrument flight rules, and you do have the VFR freedom to vary your altitude. But you've told Center one thing, so don't make changes without communicating your intended actions.

ENROUTE FREQUENCY CHANGES

You're moving along over the countryside, let's say, from Kansas City to Memphis. You've already made the initial contact with Center, as illustrated in the previous Seattle example, when you hear something like this:

> *Ctr:* *Cherokee One Four Six One Tango, contact Kansas City Center, frequency one two five point three.*

> **You:** Cherokee Six One Tango, Roger, one two five point three.

The call simply means that you're leaving that controller's sector. So you "good day" him and tune in 125.3. Then make contact on the new frequency:

> **You:** Kansas City Center, Cherokee One Four Six One Tango with you, level at seven thousand five hundred.

All you're doing is talking to a different controller at Kansas City Center who is handling the geographical sector you are now in.

Along the same line, you might hear this:

> *Ctr:* *Cherokee Six One Tango, change to my frequency* [or "contact me now on"] *one one eight point five five.*

> **You:** Roger. One one eight point five five. Cherokee Six One Tango.

Change your radio and make contact again: "Kansas City Center, Cherokee Six One Tango is with you on one one eight point five five."

Notice the difference between the calls? In the latter case, the controller said "…change to my frequency…" It's the same person you've been chatting with all along. He just wants you on a different remote site frequency closer to your present position. There's no need for the traditional "good day" or altitude report. You haven't left him. Listen for the inclusion of " me" or "my" in the controller's message. That's the tipoff.

ENROUTE ADVISORIES AND CENTER

Back to the flight itself: You're with Center and squawking the code given you. What the controller passes on to you now depends on his or her workload and the conditions along your route of flight. Typical of what you might hear are the following examples:

Example 1

Ctr: *Cherokee Six One Tango, traffic at ten o'clock, three miles, northeast bound. Altitude unknown.*

You: Negative contact. Cherokee Six One Tango. [Or, Cherokee Six One Tango has the traffic. Or, Cherokee Six One Tango requests vectors around the traffic.]

(Some old-time pilots still use the phrase "no joy" instead of "negative contact," and "tallyho" instead of "has the traffic." Neither term, however, is part of the approved phraseology.)

Example 2

Ctr: *Cherokee Six One Tango, were you advised of low-level military flights in your present area?*

You: Negative, not so advised.

Ctr: *Cherokee Six One Tango, be alert for north-south B-52 activity.*

You: Cherokee Six One Tango will be looking. Thank you.

Example 3

You see ahead of you a cloud buildup that appears to be right at your altitude. To avoid it, you decide to drop down from 7,500 to 5,500 feet. That's permissible, but advise Center ahead of time of your plans:

You: Center, Cherokee Six One Tango descending to five thousand five hundred due to clouds.

Ctr: *Cherokee Six One Tango, Roger. Report reaching five thousand five hundred.*

You: Cherokee Six One Tango wilco.

You: Center, Cherokee Six One Tango level at five thousand five hundred.

Ctr: *Cherokee Six One Tango, Roger.*

Example 4

You're still on that Kansas City–Memphis flight and are nearing Springfield, MO—a Class D airspace with Approach/Departure Control. A few miles out, Center comes on the air:

Ctr: *Cherokee One Four Six One Tango, contact Springfield Approach on one two four point niner five. Good day.*

You: Roger, Center, one two four point niner five. Thanks for your help. Cherokee Six One Tango.

A logical question: You're not landing at Springfield on this flight to Memphis, so why call Approach? The answer: Simply because you're about to enter Approach's area of radar coverage—its area of responsibility—which has a horizontal radius of about 30 miles and rises vertically to 10,000 or 12,000 feet. Once you're through the area, Approach will either hand you off to Center again (which is probable) or advise you that radar service is terminated.

Example 5

You are leaving Kansas City Center's jurisdiction and approaching that of the Memphis Center:

Ctr: *Cherokee Six One Tango, contact Memphis Center now one two four point three five.*

You: Roger. One two four point three five. Cherokee Six One Tango. Good day.

Note that Kansas City has said nothing about radar service being terminated. This means that you have been handed off to Memphis. Just change to 124.35 and make your call:

You: Memphis Center, Cherokee One Four Six One Tango is with you, level at five thousand five hundred.

Ctr: *Roger. Cherokee Six One Tango. Altimeter three zero one five. Continue present heading* [or whatever the instructions, if any, might be].

A comment or two about frequency changing is appropriate here. Let's say that you have two navcoms. The last Kansas City frequency, 125.3, is in the first navcom. When told to change to Memphis, enter 124.35 in the second navcom, but leave the first where it is. If for any reason you have to call Kansas City again, you can do so without time-consuming dial changes. Once contact is made with Memphis, merely set the first navcom to the next probable Memphis remote frequency, or turn it to any other radio aid you want.

With only one navcom aboard, be sure to write down the new frequency given you. (It's easy to forget or confuse 124.35 with 123.45—or any other combination.) If you've done this progressively from the first Departure frequency through the one or more Center frequencies, the piece of paper on your knee pad will resemble this:

KC D/C	119.0
KC Ctr	125.55
KC Ctr	125.3
MEM Ctr	124.35

Just cross out each previous frequency, but don't obliterate it. You might need it again.

Even with two navcoms, the practice of writing down each succeeding frequency makes good sense. It's a little embarrassing, after a minute or two of uncertainty or forgetfulness, to have to recontact the last controller and sheepishly ask: "What frequency did you tell me to change to?"

Example 6

As the flight progresses, you find that the headwinds are much stronger than forecast. Your forward progress is slower than the rate at which the fuel gauges are dropping, so you decide to put down at Springfield, Missouri. Because you're still in Kansas City Center's area and they have you on a nonstop flight to Memphis, a call to Center is essential:

You: Center, Cherokee Six One Tango will be landing Springfield for fuel.

Ctr: *Cherokee Six One Tango, Roger. Contact Springfield Approach on one two four point niner five.*

You: Roger. One two four point niner five. Cherokee Six One Tango. Good day.

Assuming that you filed a flight plan before leaving Kansas City, it's almost certain that the ground time at Springfield will delay your original ETA at Memphis. Accordingly, be sure to telephone the appropriate Flight Service Station and either file a new flight plan or ask that your ETA be revised. Otherwise, the FSS will start asking questions 30 minutes after that first ETA expires.

CENTER TO APPROACH

Airborne again (or maybe you didn't have to stop at all), you're nearing Memphis International. At a given point, Memphis Center will contact you with one of three possible instructions.

Example 1

Ctr: *Cherokee Six One Tango, radar service terminated. Contact Memphis Approach on one two five point eight.*

You: Roger, one two five point eight. Cherokee Six One Tango. Good day.

Center said "terminated," so you can assume that you are not being handed off. But the controller gave no instructions about changing the transponder to 1200 or any other code, which means that you leave it at the last setting—1205, or whatever it was. Then call Approach with the standard IPAI/DS:

You: Memphis Approach, Cherokee One Four Six One Tango.

App: *Cherokee One Four Six One Tango, Memphis Approach.*

You: Cherokee One Four Six One Tango is with you over Marion, level at five thousand five hundred, landing Memphis. Squawking one two zero five with Information Quebec.

Approach will take it from there.

Example 2

Ctr: *Cherokee Six One Tango, radar service terminated. Squawk one two zero zero. Contact Memphis Approach on one two five point eight.*

You: Roger, one two five point eight. Cherokee Six One Tango. Good day.

With this advice, you know there has been no handoff, so change to 1200, and give Approach the full IPAI/DS.

Example 3

Ctr: *Cherokee Six One Tango, contact Memphis Approach on one two five point eight.*

You: One two five point eight. Cherokee Six One Tango.

You: Memphis Approach, Cherokee One Four Six One Tango is with you, level at five thousand five hundred.

This was a direct handoff. Approach knew all about you and your intentions—thus no need for IPAI/DS, other than to confirm your present altitude.

Using Center as Approach/Departure Control

As mentioned in Chap. 13, Center offers limited Approach and Departure Control for many airports that would otherwise have no approach/departure services. Those so served include:

- Nontower airports that are not within radar coverage of a larger airport's Approach/Departure Control
- Tower airports where activity levels do not justify airport-based radar and which are not within the radar coverage of a larger airport's Approach/Departure Control
- Airports with a part-time tower and part-time Approach/Departure Control when those services are closed

- Tower and nontower airports within radar coverage of a larger airport's parttime Approach/Departure Control when such service is closed.

Like the other Center services discussed, VFR Approach and Departure services are available from Center on a workload-permitting basis only. Figures 14-8 through 14-10 are just three examples of airports with Center-supplied Approach and Departure services.

CONCLUSION

That's pretty much the story of Center, as far as the VFR pilot is concerned. To summarize some of the key points.

- Center exists primarily to serve IFR flights.

- Center's assistance to VFR pilots is on a workload-permitting basis.

- Use of Center by the VFR pilot is advisable but not mandatory.

- Although the VFR pilot is not under Center's control, he or she should not deviate from announced altitudes or routes of flight without advising Center in advance.

- Become familiar with the probable frequencies you will use via the ELAC and know approximately when you will be asked to change from one remote frequency to another or from one Center to another.

- Become familiar with the ELAC for more exact radio navigation and frequency-change areas.

- Write down all frequencies (planned or given) in chronological sequence so you won't forget or become confused.

WATERVILLE ROBERT LAFLEUR (WVL)   2 SW   UTC–5(–4DT)   N44°32.00' W69°40.53'   **MONTREAL**
333   B   S4   FUEL 100LL, JET A   **H–3J, L–26G**
RWY 05–23: H5500X100 (ASPH)   S–40, D–60, DT–105   HIRL   1.2% up NE   **IAP**
  RWY 05: MALSF VASI(V4L)—GA 3°TCH 51'. Trees.   RWY 23: REIL. VASI(V2L).
RWY 14–32: H2300X150 (ASPH)   S–25   MIRL
  RWY 14: Trees.
AIRPORT REMARKS: Attended 1200–0200Z‡. Numerous seagulls on and within 1½ miles of arpt. ACTIVATE HIRL Rwy
  05–23, MALSF Rwy 05—CTAF; MIRL Rwy 14–32 operates dusk–dawn. Fld conditions unavailable 2200–1300Z‡.
WEATHER DATA SOURCES: AWOS–3 120.025 (207) 877–0519.
COMMUNICATIONS: CTAF/UNICOM 122.7
  BANGOR FSS (BGR) TF 1–800–WX–BRIEF. NOTAM FILE WVL.
→ Ⓡ PORTLAND APP/DEP CON 128.35 (1100–0500Z‡)   CLNC DEL 124.6
→ Ⓡ BOSTON CENTER APP/DEP CON 128.2 (112°–292°) 124.25 (293°–111°) (0500–1100Z‡)
RADIO AIDS TO NAVIGATION: NOTAM FILE AUG.
  AUGUSTA (L) VOR/DME 111.4   AUG   Chan 51   N44°19.20' W69°47.79'   040° 13.8 NM to fld. 350/18W.
  BRACY NDB (MHW/LOM) 399   RL   N44°27.61' W69°44.09'   048° 5.1 NM to fld. NOTAM FILE WVL. NDB/LOM
    unusable byd 15 NM.
  ILS/DME 110.5   I–RLU   Chan 42   Rwy 05.   LOM BRACY NDB. Glide slope unusable byd 4° rgt side of
    LOC course. BRACY LOM/NDB unusable byd 15 NM.

Fig. 14-8. Waterville, Maine, is a nontower Class E airport that receives Approach/Departure Control from Portland or from Boston Center when the Portland Approach is closed.

AUGUSTA

BUSH FLD (AGS)   6 S   UTC−5(−4DT)   N33°22.20' W81°57.87'   **CHARLOTTE**
   145   B   S4   **FUEL** 100LL, JET A   OX 1, 2, 3,4   ARFF Index B   H−4I, 6F, L−19A, 20F
   RWY 17−35: H8001X150 (ASPH−GRVD)   S−130, D−166, DT−358   HIRL   IAP
      RWY 17: MALSR. Tree.   **RWY 35:** MALSR. Trees.
   RWY 08−26: H6001X150 (ASPH−GRVD)   S−52, D−71, DT−126   MIRL
      RWY 08: REIL. Trees.   **RWY 26:** REIL. VASI(V4L). Trees.
   AIRPORT REMARKS: Attended continuously. ACTIVATE MALSR Rwys 17 and 35—CTAF. NOTE: See Land and Hold Short
      Operations Section.
   WEATHER DATA SOURCES: ASOS (706) 790−0631. LLWAS.
   COMMUNICATIONS: CTAF 118.7   ATIS 132.75   UNICOM 122.95
      MACON FSS (MCN) TF 1−800−WX−BRIEF. NOTAM FILE AGS.
   → ® **AUGUSTA APP/DEP CON** 126.8 (170°−349°) 119.15 (350°−169°) (1145−0400Z‡)
   → ® **ATLANTA CENTER APP/DEP CON** 128.1 (0400−1145Z‡)
      AUGUSTA TOWER 118.7 (1145−0400Z‡)   **GND CON** 121.9
   AIRSPACE: CLASS D svc 1145−0400Z‡ other times **CLASS E.**
   TRSA svc ctc **APP CON**
   RADIO AIDS TO NAVIGATION: NOTAM FILE AND.
      COLLIERS (H) VORTAC 113.9   IRQ   Chan 86   N33°42.44' W82°09.72'   158° 22.5 NM to fld. 428/04W.
      EMORY NDB (HW) 385   EMR   N33°27.77' W81°59.81'   168° 5.8 NM to fld. NOTAM FILE MCN. (Unmonitored
         when twr clsd).
      BUSHE NDB (LOM) 233   AG   N33°17.22' W81°56.81'   354° 5.1 NM to fld.
      ILS 108.5   I−MZX   Rwy 17.   (ILS unmonitored when twr clsd).
      ILS 110.5   I−AGS   Rwy 35.   LOM BUSHE NDB. (ILS unmonitored when twr clsd).
      ASR (1145−0400Z‡)

Fig. 14-9. *Augusta, Georgia, has its own Approach/Departure Control, but when down, as from 0400 to 1145Z, Atlanta provides the services.*

DANBURY MUNI

DANBURY MUNI (DXR)   3 SW   UTC−5(−4DT)   N41°22.29' W73°28.93'   **NEW YORK**
   458   B   S4   **FUEL** 100LL, JET A   OX 4   TPA—See Remarks   ARFF Index Ltd.   L−25B, 28H
   RWY 08−26: H4422X150 (ASPH−GRVD)   S−38, D−70   MIRL   IAP
      RWY 08: REIL. Thld dsplcd 368'. Trees.   **RWY 26:** REIL. Thld dsplcd 734'. Tree.
   RWY 17−35: H3135X100 (ASPH)   S−50, D−65
      RWY 17: Thld dsplcd 223'. Pole.   **RWY 35:** Thld dsplcd 231'. Tree.
   AIRPORT REMARKS: Attended 1200Z‡−dusk. Deer and birds on and invof arpt. Rwy 17−35 CLOSED ngts and to air
      carrier ops with more than 30 passenger seats. Rwy 08 CLOSED to landings for air carrier ops with more than
      30 passenger seats. ACTIVATE MIRL Rwy 08−26 and REIL Rwys 08 and 26 when twr clsd−119.4. Acft using
      Rwy 35 not visible from twr descending below 1300 ft on base leg until approaching ½ mi final due to natural
      terrain. TPA—1701(1243)—Jet acft 2201(1743). PPR 24 hours for unscheduled air carrier operations with more
      than 30 passenger seats; call arpt manager 203−797−4624. PPR for formation tkf/ldg; ctc arpt manager.
      Rotating beacon located one−half mile S of arpt on top of a hill.
   WEATHER DATA SOURCES: LAWRS.
   COMMUNICATIONS: CTAF 119.4   ATIS 127.75 (1200−0300Z‡)   UNICOM 122.95
      BRIDGEPORT FSS (BDR) TF 1−800−WX−BRIEF. NOTAM FILE DXR.
   → ® **NEW YORK APP/DEP CON** 126.4   **CLNC DEL** 128.6 (When DXR twr clsd.)
   →   **TOWER** 119.4 (1200−0300Z‡)   **GND CON** 121.6
   AIRSPACE: CLASS D svc 1200−0300Z‡ other times **CLASS G.**
   RADIO AIDS TO NAVIGATION: NOTAM FILE ISP.
      CARMEL (L) VORW/DME 116.6   CMK   Chan 113   N41°16.80' W73°34.88'   051° 7.1 NM to fld. 690/12W.
      ILS/DME 111.55   I−DXR   Chan 52Y   Rwy 08.   LOC unmonitored when twr clsd. DME unusable byd 3°
         right side of course; byd 10° left side of course and byd 9.5 NM.

Fig. 14-10. *Danbury, Connecticut, has its own tower but no local Approach/Departure Control facility. Hence New York provides this service, even when the Danbury tower is open.*

- Plan your communication to Center before picking up the mike.

- Rehearse your message so that you sound like a pro—not a hesitant amateur.

- Center is one more eye in the sky to help you get where you want to go. I urge that you use this source, when possible, for that added insurance and that you use it on every cross-country flight. Too many non-IFR pilots, because of lack of knowledge and confidence, are afraid of it—when there is no reason for fear of any sort. If the controller can't accommodate you, he or she will tell you; if you don't understand a direction, ask him to repeat or clarify. I've heard airline pilots, who are presumably pros, do this on any number of occasions. In every case, the controller honored the request without rancor or intimidation. He'll do the same for you, if you sound as though you know what you're doing and what's going on.

One final observation: there's an old saying that "what we're not up on, we're down on." What we don't understand, we're either against or we fear. Uncertainty breeds insecurity. This chapter (and, in fact, the whole book) tries to provide some level of knowledge that will lead to understanding and greater pilot confidence in the field of radio communications.

Rarely can examples and the written word accomplish the entire task, however. Consequently, I recommend that you visit the Center nearest you. Call a supervisor, explain what you want, and request a brief tour of the facility. See what's going on. Talk to a couple of controllers. Listen to the communications between ground and air. Study the screens and the blips that identify the various aircraft.

Firsthand experience is the best way I know to bring meaning to words and examples to life. The FAA urges all pilots to visit its facilities—Center, Tower, Flight Service, and the rest. The facilities are glad to welcome you, and, workload permitting, will see that your tour is complete and educational. It's worth a couple of hours of your time to put to rest any feelings of uncertainty. "What we're up on, we're not down on." The saying works in reverse as well.

15

In the Event of Radio Failure

DESPITE THE SOPHISTICATION OF MODERN AVIONICS, THINGS CAN go wrong. When your radio decides to take a rest, it is comforting to know what you need to do to get your airborne vehicle safely back on the ground.

As *AIM* says, it's virtually impossible to establish fixed procedures for all situations involving two-way radio failure. Preferred actions, however, can be outlined. Within those parameters, every pilot should have a plan in the unlikely event communications are lost. Just as the intelligent pilot is mentally prepared for an engine failure, so is he or she ready to cope with a radio failure.

ONE PREVENTIVE STEP

It's pretty hard to determine the potential for radio failure during the preflight check. Of course, a loose or broken antenna is an obvious signal, as is a navcom that slips in and out of its housing rack, a circuit breaker that has popped, or a frayed magneto wire. Otherwise, there's not a lot you can check.

One symptom of a potential problem, however, is the alternator belt. If you can inspect it visually, check it for wear and frays. Whether you can see it or not, check its tension. If it's loose, have it replaced or tightened. Not all radio failures can be traced to alternators or alternator belts, but the malfunctioning of one or the other will result in a general power loss, with the battery supplying what little remaining electrical juice it can generate. And the battery won't last forever under those conditions.

WHEN A FAILURE IS SUSPECTED

Every once in a while, the radio might give off an eerie silence; what has been a pattern of constant chatter suddenly is no more; or your calls to a ground facility seem to fall on deaf ears; whatever the case, you begin to wonder if....

Before getting too concerned, adjust the squelch or turn up the volume on the radio. If you get the characteristic static, the radio is working. It's just that there's been an unusual dearth of communication activity over the past few minutes. If there is no static sound, push the set in a little. Vibration might have caused it to slip out of its rack just enough to break electrical contact. Should you appear not to be transmitting, check all connections. Is the mike plugged in? If you are using a headset, is it plugged in? Little things can happen in normal operations that produce the symptoms of radio failure, but they can be corrected with only a minor adjustment.

However, when a previously clear reception starts to break up, or you develop an unusual hum in the speaker, trouble might be brewing. Check the ammeter. Is it still showing a charge? Check the circuit breakers. If a navcom button has popped, let it cool off for a couple of minutes before resetting it. Does that do the trick? If so, you're probably all right...for a while. However, something is shorting out and should be checked as soon as you're back on the ground.

But suppose the ammeter shows no charge at all. Test it again by turning on the landing light. If the needle doesn't move, you can be sure the alternator has died or its belt has broken. In either case, what electrical power remains is coming from the battery alone. This being the case, turn off *all* nonessential electrical equipment, except one radio (assuming you can get some reception over it), and head for home or the nearest airport.

Unless you have experienced a radio or electrical failure in flight, the ammeter is probably one of your least monitored instruments. I suggest that you permanently incorporate it in the panel-scanning process. The sooner a problem is caught, the better, because the life of a battery, once the alternator is gone, is about two hours. After that, you'll have no electrical power for lights, navcoms, transponder, gauges, and the like. The engine won't stop, because the ignition system is independent of the alternator-battery system—but a functioning engine, as critical as that is, is about all you'll have left.

WHEN A FAILURE IS CONFIRMED

There is no question: The failure is real. The reception is getting weaker and weaker and your transmission is almost unreadable. To explain your options, let's set up five VFR situations—four in the air and one on the ground—that you might someday encounter.

Case 1

This example is the simplest of all. You're flying locally out of nontower Class G or E airport, or you're on a short cross-country flight with no intermediate landing planned. Because two-way communications are not required under these conditions, there's no need to alert ATC with the standard radio failure transponder codes when the set goes bad. Just continue to squawk 1200, stay away from Class B, C, and D airspaces, and land at your own or some other multicom or unicom field to get the radio fixed. Do listen, though, for CTAF traffic advisories, if the receiver works at all. And do be careful as you enter the traffic pattern. You're coming in unannounced and you might not know what other traffic is in the pattern or where it is—both of which are conditions that open the door to trouble.

Case 2

This time your destination is the primary airport in a Class B. The symptoms of a sick radio have been plaguing you for the past hour or so, but there's enough juice left to monitor the ATIS, contact Approach Control, announce your intentions, and get clearance into the airspace. Once inside, however, the radio becomes useless. You can't decipher what Approach is communicating and Approach can't understand your transmissions. Now what do you do?

First, squawk 7600.

Second, continue through the airspace to the airport environment, Approach knows what you intended to do and will protect you just as it would an IFR aircraft.

Third, knowing from the ATIS the active runway, winds, and so on, enter the airport area with the landing pattern in mind. As you do, however (and this is most important), watch the tower for light gun signals that will clear you to land or tell you to keep circling. (These signals are discussed further in Case 3 and are illustrated in Fig. 15-1.) When you see the green light, that's your landing clearance. Once on the ground and off the active runway, continue to monitor the tower for the flashing green light that clears you to taxi.

Now get the radio fixed. The odds are almost certain that you won't be allowed to depart the Class B airport until two-way communications are reestablished.

The above scenario applies equally:

- To radio failure in Class C airspace

- If your destination is a satellite airport that underlies but is not in a Class B or Class C and you are already in either of the airspaces

- If you're in the process of simply transiting a Class B or C airspace

In essence, once you've made contact with Approach and have been cleared into the Class B or Class C airspace, keep right on going in accordance with your announced intentions. That's much safer than wandering around in a sort of panic because you've lost radio contact with the ground. When you deviate from what you've told the controller, he or she doesn't know what to expect—which could cause a fair amount of confusion in his or her efforts to maintain an orderly flow of traffic.

	MEANING		
COLOR AND TYPE OF SIGNAL	MOVEMENT OF VEHICLES EQUIPMENT AND PERSONNEL	AIRCRAFT ON THE GROUND	AIRCRAFT IN FLIGHT
Steady green	Cleared to cross, proceed, or go	Cleared for takeoff	Cleared to land
Flashing green	Not applicable	Cleared for taxi	Return for landing (to be followed by steady green at the proper time)
Steady red	STOP	STOP	Give way to other aircraft and continue circling
Flashing red	Clear the taxiway/runway	Taxi clear of the runway in use	Airport unsafe, do not land
Flashing white	Return to starting point on airport	Return to starting point on airport	Not applicable
Alternating red and green	Exercise extreme caution	Exercise extreme caution	Exercise extreme caution

Fig. 15-1. *The meanings of the various control tower light gun signals. Watch out for them, especially in a radio failure situation.*

Case 3

This time you want to land at a Class D tower-controlled airport that has no radar, has no local Approach Control facility, and is miles away from either a Class B or C airspace. The radio is dead, but you squawk the usual 7600 RF code. Considering the circumstances, of what value is that squawk, and what do you do now?

Well, the nearest Center (or its remoted outlet) or Approach, seeing the radio-failure (RF) code flashing on its screen, will start tracking you on radar. If it then appears that you're heading for the Class D airport, one of the facilities may notify the tower by phone of your probable landing intentions.

As you near the airport, try to establish contact. Just possibly there's enough life in the battery or the radio to get the necessary pattern instructions and landing clearance. If not, cross over the field at least 500 feet above pattern altitude, observe the flow of traffic and the active runway, and be very much on the lookout for other aircraft.

While still well above pattern altitude, fly upwind over the active runway, the purpose being to attract the controller's attention. If Center or Approach has alerted him or her to the arrival of an RF aircraft, he or she will be watching and will try to reach you by radio. Should the radio be working at all, you'll hear something like: "Aircraft over the runway, Albany Tower. If you receive me, rock your wings." This you do, followed by: "Roger, do you intend to land at Albany?" Again rock the wings. The controller will then fit you into the pattern and give you landing clearance.

If the radio is completely dead, you might have to fly over the runway a couple of times, but be sure to keep an eye on the tower. The controller, knowing that he or she is not in radio communication, will instruct you via the light gun by aiming the gun at your aircraft and flashing one of the signals listed in Fig. 15-1—most probably a steady green or a steady red.

The only problem about seeing the light is the location of the tower in relation to the landing runway. You might have difficulty seeing it if you're in the left seat and the tower is on your right. Should that be the case, watch for the light on the crosswind or down-wind leg—still 500 feet or so above pattern altitude. Once you've got the steady green, rock your wings in acknowledgment.

Don't fail, though, to check the tower frequently for any subsequent visual instructions. Something might happen, especially on the final approach, that would cause the tower to want you to abort your landing—an unauthorized aircraft or a ground vehicle on the runway, you forgot to lower your gear, a sudden accident, a jet closing rapidly behind you, an aircraft cutting in ahead of you, the spacing between aircraft is too close, or whatever. With no radio, the gun is the only means of communication the tower has with you, so be alert to the fact that you might get a steady or flashing red light any time in the pattern.

CASE 4

You want to land at a primary Class B or C airport. You have not been using Center and the radio fails before you have contacted Approach. Your only alternative in this situation is to land somewhere outside of the B or C airspace and then request entry approval from Approach by telephone. Although highly unlikely without two-way radio communications, Approach might grant permission if the traffic is light and you are within a few miles of the primary field.

CASE 5

You're on the ground at some uncontrolled field, perhaps your home base. The radio is dead and you want to fly to a Class D tower-controlled airport, which either underlies or is far from a Class B or C airspace, for repairs. What are your options?

Really, you have only one: A telephone call to the tower supervisor before departing. In the call, explain your predicament, estimate the time of arrival in the area, describe your aircraft, and request approval to enter the ATA. If the supervisor can grant your request, he or she will probably give you instructions about pattern altitudes and what to do when you get into the Class D airspace.

Since you're already on the ground, there's no emergency. Permission to enter the airspace is thus a matter of option, largely depending on probable traffic volume, weather conditions, field conditions, or other factors that might affect approval or denial of the request. It's reasonable to assume that the tower will do what it can to help you, but that assistance isn't automatic. Two-way radio communications are too critical in a busy terminal environment—which means that you might have to look elsewhere for a repair shop at some nontower field.

CONCLUSION

The preceding recommendations, suggestions, or procedures, are fine; they will *almost* always get you down without any problems. But they don't work 100 percent of the time.

I knew a pilot who once lost an alternator belt in the vicinity of a Class B airspace while waiting to land at an underlying Class D radar-equipped airport. He could receive, although the reception was scratchy and getting worse, but not transmit. Tuned to the tower and squawking the RF code produced no response, so he flew up the Class D landing runway at 2,000 feet agl.

Midway over the airport, he heard: "Aircraft over the runway, rock your wings if you receive me and are landing at [*name of airport*]."

Replying affirmatively, the pilot was sequenced into the pattern and cleared to land.

Once off the runway, he switched to ground, but if any taxi clearance was transmitted, the radio was too far gone for him to know. And there was no light from the tower.

After a couple of futile minutes, he taxied slowly to the ramp and placed a telephone call to the tower to explain his unauthorized taxiing and to ask if the RF code had been received. The response was no-because the radar was down. Why no light signals? Sorry about that, but "our gun is broken."

So, not all malfunctions involve the aircraft. In this case, it was the alertness of the controller who spotted him above the runway and led him down to an uneventful landing at a busy airport.

Radio failures are relatively rare, but when they do occur, you should know what to do. The best advice: Don't bust into a controlled airport without a radio unless you have no safe alternative. Get out of the vicinity, find a nontowered field, and have the problem repaired (or make a phone call to the tower to request a no-radio clearance).

16
A Cross-Country Flight: Putting It All Together

THE PRECEDING CHAPTERS HAVE TAKEN YOU FROM MULTICOM through Center. Now let's put it all together with a mythical cross-country trip. Except for multicom airports, I've tried to incorporate at least one example of the radio communications with each of the various facilities I've been discussing. I hope, then, the "flight" will serve as a reasonable model for the real-life excursions you might make. Naturally, your itineraries will differ, as will the frequencies, but the principles illustrated should not vary because of that.

The flight will take you from the Kansas City Downtown Airport to Omaha's Eppley Field, where you'll pick up two friends. From there, you'll go to the Minneapolis International Airport. After completing your business in the Twin Cities, you'll drop off one of your friends at Mason City, Iowa, then the other friend at Newton, Iowa. From Newton, it's nonstop back to Kansas City.

The entire flight presupposes that you have a Mode C transponder, two navcoms, distance measuring equipment (DME), and no loran or GPS. If you don't have the luxury of multiple navcoms, the principles are the same, but frequency-changing is a little less convenient.

Chapter Sixteen

THE FLIGHT ROUTE

Without loran, you decide to fly the airways whenever possible, even though doing so will add a few miles to the trip. Accordingly, the route of flight, VORS, course headings, and point-to-point nautical mileage resemble Fig. 16-1.

As 900 statute miles is obviously too much territory to cover in one day, with stops and business enroute, you plan to stay overnight in Minneapolis. You'll also refuel at

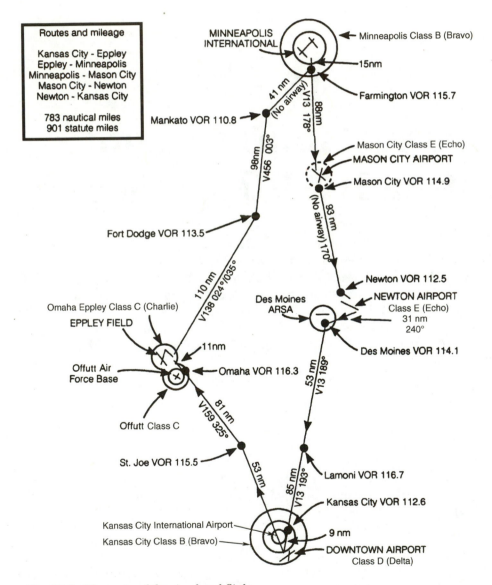

Fig. 16-1. *The route of the simulated flight.*

each stop, except Mason City, and use the appropriate Air Route Traffic Control Centers for VFR advisories.

The reason for this itinerary is simply to review the communication procedures involved in the various situations, airports, and controlling agencies we discussed in the previous chapters. To wit:

- Kansas City Downtown Airport, a Class D airport, lies under a Class B airspace.
- Omaha Eppley is a Class C airspace.
- Minneapolis International Airport is a Class B airport.
- Mason City is a unicom Class E airport with a weather observer, but no tower or FSS is on the field.
- Newton is a Class E uncontrolled field with only unicom.

To go through the entire flight planning process is beyond the scope of this book. Let's assume that all preliminaries have been completed, including weight-and-balance calculations and filing the flight plan.

RECORDING THE FREQUENCIES

As part of the preflight planning, write down in sequence the known or probable frequencies you will use. Some, particularly Center's, might differ from what you expect, but at least you'll be ready for the majority that come into play.

I suggest you do not record all the frequencies on one piece of paper for a flight like this with five different legs. Enter the frequencies to Omaha on one sheet, those from Omaha to Minneapolis on another, and so on (Figs. 16-2 through 16-6). Then number each sheet in sequence. As you leave one frequency and progress to the next, draw a line through (but don't obliterate) the one you have just left. This preliminary recording and progressive deleting provides cockpit organization and minimizes some of the confusion that is often the bane of the private pilot.

Two other suggestions: On each segment page, provide space for the ATIS information at the destination airport. When you're on the ground before departure, this isn't so important because you can listen to the local information as many times as necessary. In the air, however, it's another matter. Center has handed you off to Approach, but before contacting Approach, you should have monitored the ATIS—which means that you don't have a lot of time to absorb the data being transmitted. Approach is expecting to hear from you rather promptly after the handoff.

Consequently, to expedite matters, line out a box on the flight segment page so that you can record the crucial information: (1) the phonetic alphabet; (2) the sky or ceiling; (3) visibility; (4) temperature; (5) dewpoint; (6) altimeter setting; (7) runway in use; and (8) other important information that might be included. Now, when you call Approach, you can advise the controller that you "have Charlie," or whatever, and be sure that you have it accurately. Memories do fail us.

Second, sketch in on the same segment page a rough diagram of the destination airport runways, including the distance and direction from town, field elevation, and pattern

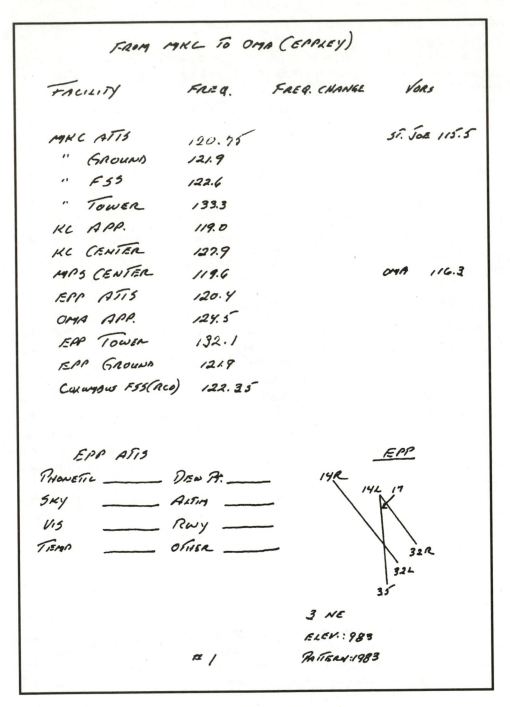

FROM MKC TO OMA (EPPLEY)

FACILITY	FREQ.	FREQ. CHANGE	VORS
MKC ATIS	120.75		ST. JOE 115.5
" GROUND	121.9		
" FSS	122.6		
" TOWER	133.3		
KC APP.	119.0		
KC CENTER	127.9		
MPS CENTER	119.6		OMA 116.3
EPP ATIS	120.4		
OMA APP.	124.5		
EPP TOWER	132.1		
EPP GROUND	121.9		
Columbus FSS (RCO)	122.35		

EPP ATIS

PHONETIC _____ DEW PT. _____
SKY _____ ALTIM _____
VIS _____ RWY _____
TEMP _____ OTHER _____

EPP

14R
14L 17
32R
32L
35

3 NE
ELEV: 983
PATTERN: 1983

1

Fig. 16-2. *The fight planning notes from MKC to OMA (Eppley).*

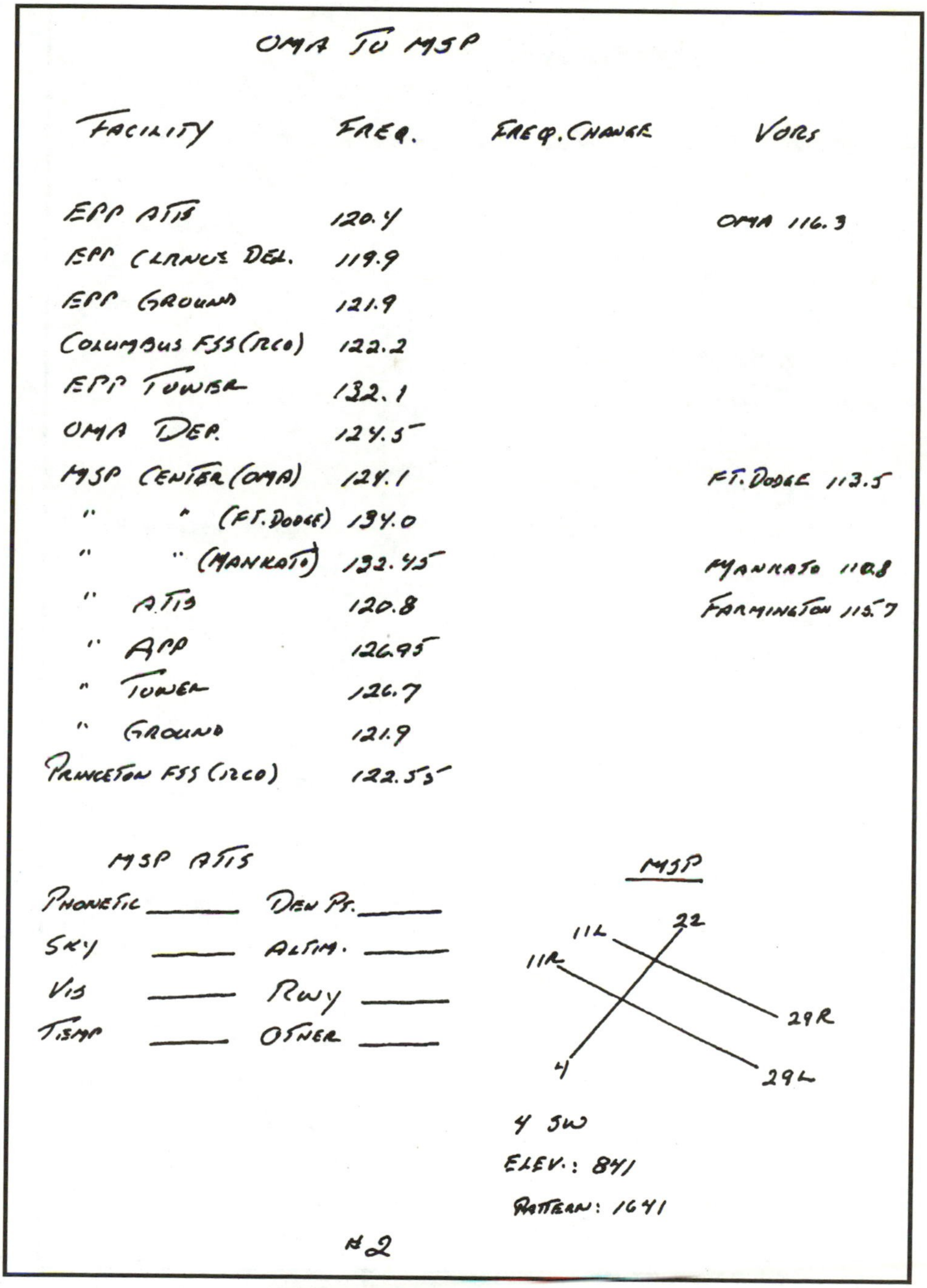

OMA TO MSP

FACILITY	FREQ.	FREQ. CHANGE	VORS
EPP ATIS	120.4		OMA 116.3
EPP CLRNCE DEL.	119.9		
EPP GROUND	121.9		
COLUMBUS FSS (RCO)	122.2		
EPP TOWER	132.1		
OMA DEP.	124.5		
MSP CENTER (OMA)	124.1		FT. DODGE 113.5
" " (FT. DODGE)	134.0		
" " (MANKATO)	132.45		MANKATO 110.8
" ATIS	120.8		FARMINGTON 115.7
" APP	126.95		
" TOWER	126.7		
" GROUND	121.9		
PRINCETON FSS (RCO)	122.55		

MSP ATIS

PHONETIC _____   DEW PT. _____
SKY _____   ALTIM. _____
VIS _____   RWY _____
TEMP _____   OTHER _____

MSP

4 SW
ELEV.: 841
PATTERN: 1641

#2

Fig. 16-3. The flight planning notes from OMA to MSP.

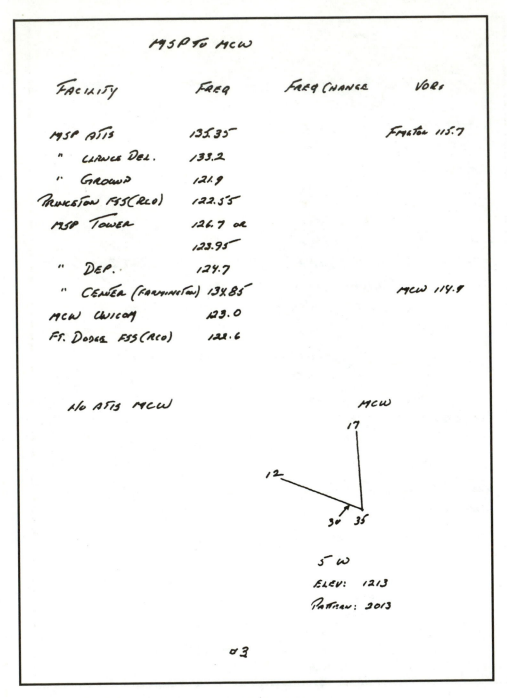

MSP TO MCW

FACILITY	FREQ	FREQ CHANGE	VORs
MSP ATIS	135.35		Fargton 115.7
" CLRNCE DEL.	133.2		
" GROUND	121.9		
PRINCETON FSS (RCO)	122.55		
MSP TOWER	126.7 or		
	123.95		
" DEP.	124.7		
" CENTER (FARMINGTON)	134.85		MCW 114.9
MCW CNICOM	123.0		
FT. DODGE FSS (RCO)	128.6		

NO ATIS MCW

MCW

17

12

30 35

5 W

ELEV: 1213

PATTERN: 2013

O 3

Fig. 16-4. The flight planning notes from MSP to MCW.

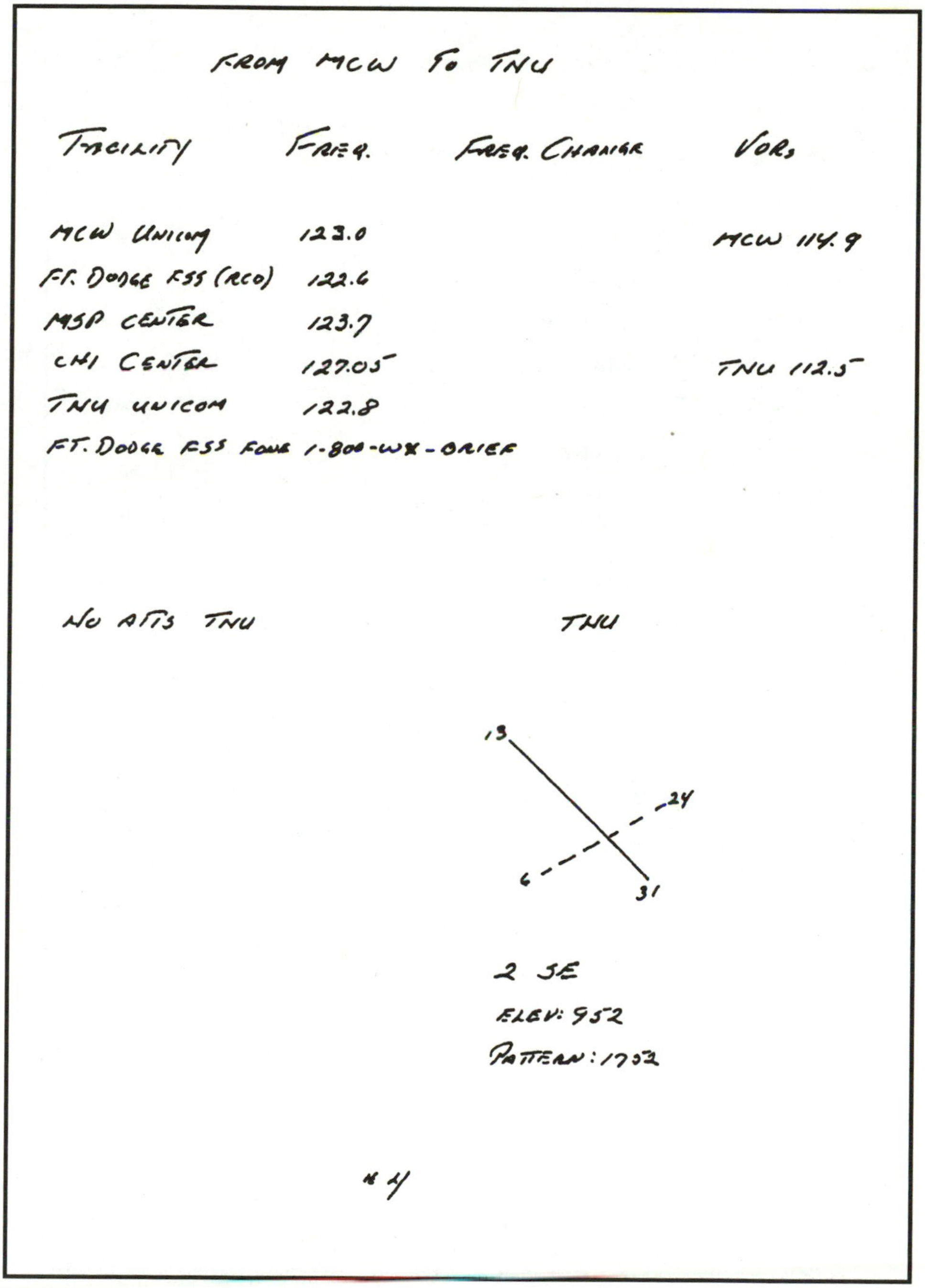

FROM MCW TO TNU

FACILITY	FREQ.	FREQ. CHANGE	VORs
MCW UNICOM	123.0		MCW 114.9
FT. DODGE FSS (RCO)	122.6		
MSP CENTER	123.7		
CHI CENTER	127.05		TNU 112.5
TNU UNICOM	122.8		
FT. DODGE FSS FONE	1-800-WX-BRIEF		

NO ATIS TNU   TNU

13
24
6
31

2 SE
ELEV: 952
PATTERN: 1752

N 4

Fig. 16-5. The flight planning notes from MCW to TNU.

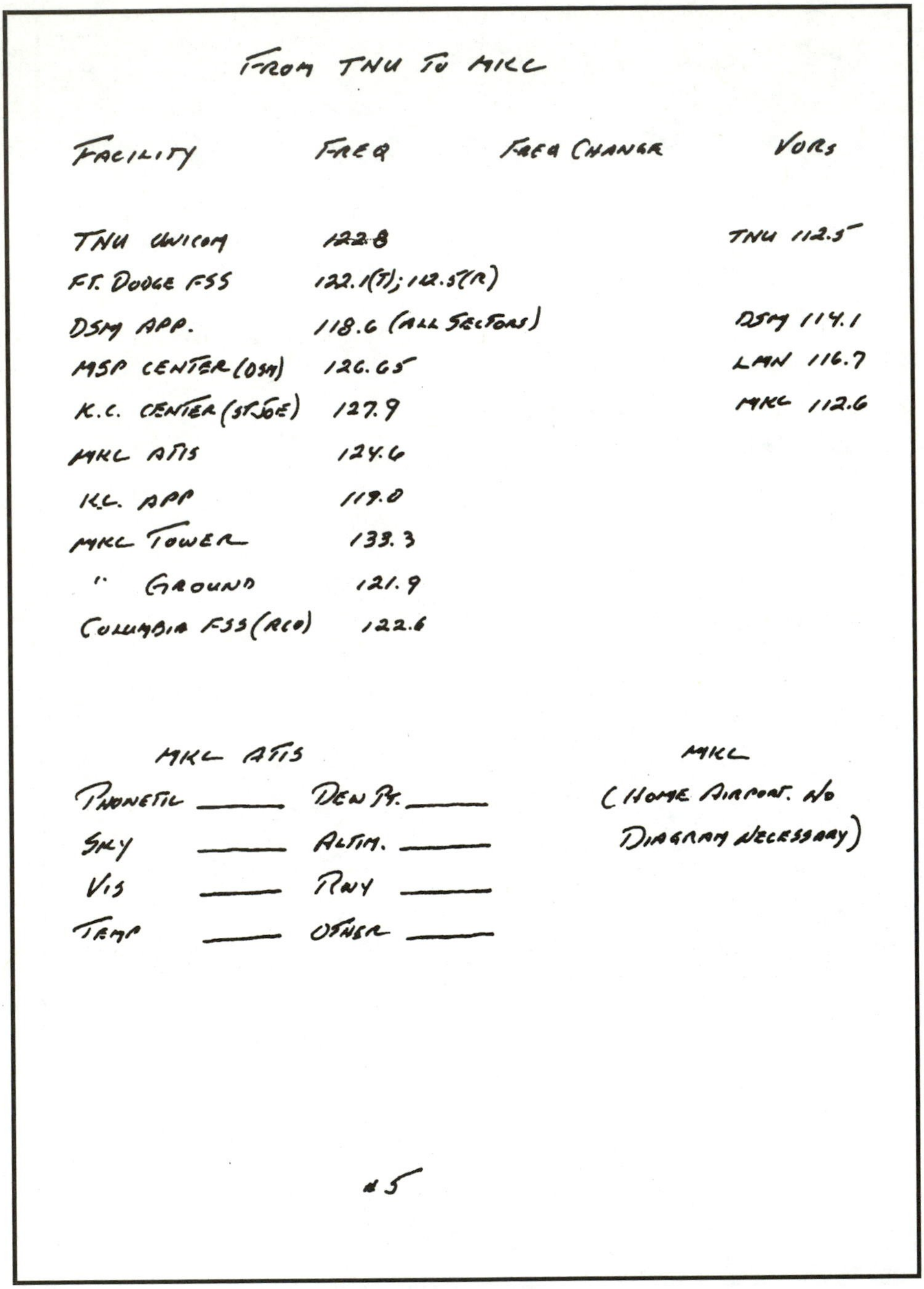

FROM TNU TO MKC

FACILITY	FREQ	FREQ CHANGE	VORs
TNU UNICOM	122.8		TNU 112.5
FT. DODGE FSS	122.1(T); 112.5(R)		
DSM APP.	118.6 (ALL SECTORS)		DSM 114.1
MSP CENTER (DSM)	126.65		LMN 116.7
K.C. CENTER (ST. JOE)	127.9		MKC 112.6
MKC ATIS	124.6		
K.C. APP	119.0		
MKC TOWER	133.3		
" GROUND	121.9		
COLUMBIA FSS (RCO)	122.6		

MKC ATIS

PHONETIC ____	DEW PT. ____
SKY ____	ALTIM. ____
VIS ____	RNY ____
TEMP ____	OTHER ____

MKC
(HOME AIRPORT. NO
DIAGRAM NECESSARY)

♯5

Fig. 16-6. *The flight planning notes from TNU to MKC.*

altitude. If you want, you can add the taxiways and building locations. *AOPA's Aviation USA,* published by the Aircraft Owners and Pilots Association, provides diagrams and data of more than 13,000 airports in the United States and its possessions. It's an excellent source for determining the layout and runway data of whatever airport you have in mind. Similar diagrams are found on state aeronautical charts and on instrument approach charts.

The purpose of the sketch is apparent: it minimizes mental or spatial confusion when going into a strange airport for the first time. Just as important, it can reduce the number of questions or inquiries you might have to make of Approach, the Tower, or Ground Control.

The examples (Figs. 16-2 through 16-6) aren't very fancy, but that's not the point. Practicality is the objective. (Just don't rely on the frequencies cited as being current. They do change.)

THE FLIGHT AND RADIO CONTACTS

Equipped with the necessary charts—sectional charts as well as the Enroute Low Altitude charts (ELAC)—the frequencies recorded, the flight planned, and the flight plan filed, you're ready to depart.

Kansas City to Omaha

With the engine started, the first order of business is to tune to the Downtown Airport ATIS on 120.75. Put this in the first radio and set up Ground Control, 121.9, on the second.

ATIS: *This is Kansas City Downtown Airport Information Delta. One six four five Zulu weather. Five thousand scattered, measured ceiling ten thousand broken, visibility eight. Temperature seven eight, dewpoint five five, wind one seven zero at ten, altimeter two niner niner eight. ILS Runway One Niner in use, land and depart Runway One Niner. Advise you have Delta.*

Now change the first radio to the FSS frequency of 122.6, which is the frequency that Flight Service gave you to open your flight plan. Put the transponder on STANDBY and call Ground Control on the second radio:

You: Downtown Ground, Cherokee One Four Six One Tango at Hangar 6, VFR Omaha with Information Delta.

GC: *Cherokee One Four Six One Tango, taxi to Runway One Niner.*

You: Roger. Taxi to One Niner, Cherokee Six One Tango.

Stay on the Ground frequency. Clearance to taxi doesn't mean that the controller might not have subsequent instructions for you:

GC: *Cherokee Six One Tango, give way to the Baron taxiing south.*

You: Wilco, Cherokee Six One Tango.

After completing the pre-takeoff check, call Ground and advise them that you're leaving the frequency momentarily to go to Flight Service. Now change radio 2 to the tower frequency of 133.3, switch to radio 1, which is already tuned to Flight Service, and open the flight plan:

You: Columbia Radio, Cherokee One Four Six One Tango on one two two point six, Kansas City.

FSS: *Cherokee One Four Six One Tango, Columbia Radio.*

You: Would you please open my VFR flight plan to Omaha Eppley at one three three five UTC?

FSS: *Cherokee Six One Tango, Roger We'll open your flight plan at one three three five UTC.*

You: Roger, thank you. Cherokee Six One Tango.

Two points to remember:

1. Be sure to add five to ten minutes to your expected takeoff time in case your departure is delayed.
2. Opening a flight plan while still on the ground is possible only when there is an FSS, RCO, or VOR voice facility on the field.

As your course to Omaha is northwesterly, the most direct routing is through the Kansas City Class B airspace, so change radio 1 to Approach Control on 119.0. (Remember that you're under the Class B airspace and must contact Approach to enter it.) With radio 1 set up, the next call is to the tower on radio 2:

You: Downtown Tower, Cherokee One Four Six One Tango ready for takeoff with north departure.

Twr: *Cherokee One Four Six One Tango, hold short. Landing traffic.*

You: Cherokee Six One Tango holding short.

Twr: *Cherokee Six One Tango, cleared for takeoff. Left turn for north departure approved. Remain clear of the final approach course. Contact Approach when airborne.*

You: Will do. Cherokee Six One Tango. [Now switch the transponder to ALT.]

Even if the tower has advised you to contact Approach after takeoff, it doesn't hurt to request the frequency-change approval or to inform the tower that you're about to make the change. The controller might have reasons for wanting you to stay with him for a few minutes. The next call then would be:

You: Tower, Cherokee Six One Tango requests frequency change [or "going to Approach"].

Twr: *Cherokee Six One Tango. Frequency change approved.*

You: Cherokee Six One Tango. Good day.

Go now to radio 1 and call Approach on 119.0:

You: Kansas City Approach, Cherokee One Four Six One Tango.

App: *Cherokee One Four Six One Tango, Kansas City Approach, go ahead.*

You: Cherokee One Four Six One Tango is off Downtown at two thousand, requesting six thousand five hundred to Omaha, and would like clearance through the Bravo airspace.

App: *Cherokee Six One Tango, squawk zero two five two and ident. Remain clear of the Bravo airspace.*

You: Cherokee Six One Tango squawking zero two five two. Remaining clear.

Remember that you have not yet been cleared into the airspace, so stay below this Class B's 2,400-foot floor until you hear the next message:

App: *Cherokee Six One Tango, radar contact. Cleared to enter the Class B airspace. Fly heading three four zero and maintain four thousand five hundred.*

You: Roger. Cleared to enter the Class B airspace, heading three four zero, leaving two thousand for four thousand five hundred. Cherokee Six One Tango.

As you start your turn and begin the climb, you might hear this before actually entering the Bravo airspace.

App: *Cherokee Six One Tango, traffic at twelve o'clock, three miles, southbound. Unverified altitude two thousand niner hundred.*

You: Cherokee Six One Tango looking.

And then:

App: *Cherokee Six One Tango, traffic no longer a factor.*

You: Roger. Cherokee Six One Tango.

You: Cherokee Six One Tango, level at four thousand five hundred.

App: *Cherokee Six One Tango, Roger. Climb and maintain six thousand five hundred.*

You: Out of four thousand five hundred for six thousand five hundred. Cherokee Six One Tango.

You: Cherokee Six One Tango level at six thousand five hundred.

> **App:** *Cherokee Six One Tango, Roger. Turn left heading three one zero. Proceed direct St. Joe when able.*

> **You:** Roger. Three one zero on the heading. Direct St. Joe when able. Cherokee Six One Tango.

The St. Joe VOR lies about 50 nm north of Kansas City. Approach is saying that you are to tune the nav receiver to the VOR frequency of 115.5. When you have the altitude and distance to get a steady needle reading, you're cleared to proceed directly on course to St. Joe.

Other instructions or traffic alerts might follow. While you have the time, however, you should be setting up radio 2 (until now still on the tower frequency) to Kansas City Center, which you expect to be 127.9. Approach might give you a different frequency, but if not, you're ready to contact Center without delay.

As you work your way northward toward St. Joe, you reach the Class B limits:

> **App:** *Cherokee Six One Tango, position 30 miles northwest of International, departing the Bravo airspace. Radar service terminated. Squawk one two zero zero. Frequency change approved.*

> **You:** Cherokee Six One Tango. Can you hand us off to Center?

> **App:** *Unable at this time, Six One Tango. Contact Center on one two seven point niner. Good day.*

> **You:** One two zero zero and one two seven point niner. Roger. Cherokee Six One Tango.

There is no handoff, so the call to Center requires the full IPAI/DS:

> **You:** Kansas City Center, Cherokee One Four Six One Tango.

> **Ctr:** *Cherokee One Four Six One Tango, Kansas City Center, go ahead.*

> **You:** Cherokee One Four Six One Tango is two zero south of St. Joe VOR, level at six thousand five hundred enroute Omaha Eppley. Squawking one two zero zero. Request VFR advisories, if possible.

> **Ctr:** *Cherokee Six One Tango, squawk four one five zero and ident.*

> **You:** Cherokee Six One Tango squawking four one five zero.

> **Ctr:** *Cherokee Six One Tango. Radar contact. St. Joe altimeter two niner niner eight.*

> **You:** Roger. Cherokee Six One Tango.

What comes over the air now depends primarily on the traffic along your line of flight. You might be told to assume a different heading; you might be alerted to the proximity of other aircraft; you might be alerted to military training flights. Or you might hear nothing.

Regardless of the communiqués, or lack thereof, you have some navigating to do, along with keeping an ear out for your call sign. You're still on course to the St. Joe VOR, with the VOR head needle centered and the DME recording the distance to the VOR, the time to the station, and your current ground speed.

As soon as you pass over St. Joe and get a "FROM" reading on the VOR, turn to a heading of 325 degrees. Once the needle has centered itself, indicating that you're now on V159, stay tuned to that station for another 35 or 40 miles. If you have a second nav-com, you can tune in the Omaha VOR on 116.3 at any time, but at your altitude you probably won't get a very reliable "TO" indication until you're within a 60- or 70-mile range of the station. So maintain the outbound course from St. Joe with the "FROM" reading on one VOR, and then when the other VOR needle settles down to a steady centered position with a "TO" reading, rely on it to lead you to the Omaha station.

Meanwhile, don't get too enthralled with VOR-to-VOR navigating alone. Something might go wrong, so it's always wise to check your progress against the route you've laid out on the sectional chart. If you should lose the nav portion of the radio, or if the VOR suddenly had a mechanical failure, it could be rather important to know where you were along the route. Things electronic are great, but they're not infallible.

Eventually, as you move down V159, Kansas City Center comes on the air:

Ctr: *Cherokee Six One Tango, contact Minneapolis Center now on one one niner point six. Good day.*

You: One one niner point six. Thank you for your help. Cherokee Six One Tango.

No comment here about radar service being terminated, so this is a handoff by Kansas City to Minneapolis. If you've already changed radio 1 from Kansas City Approach to Minneapolis, all you have to do now is go back to radio 1 and introduce yourself:

You: Minneapolis Center, Cherokee One Four Six One Tango with you, level at six thousand five hundred.

Ctr: *Cherokee Six One Tango, Roger. Altimeter three zero zero two.*

Unless you are told otherwise, maintain your present heading and altitude. Again, there might or might not be instructions or advice for you. Eventually, however, as you near the Omaha Class C Charlie airspace, Center will come back on:

Ctr: *Cherokee Six One Tango, position 30 miles south of Eppley. Contact Omaha Approach on one two four point five. Good day.*

You: Roger. One two four point five. Cherokee Six One Tango. Good day.

This, too, is a handoff, so tune in to Approach. Before making the call, however, if you haven't done so already, get the ATIS on 120.4 for the current Eppley information. Now you're ready to contact Approach:

You: Omaha Approach, Cherokee One Four Six One Tango is with you, level at six thousand five hundred with Information Echo.

> *App: Cherokee Six One Tango, Roger. Maintain present heading and altitude.*

As you enter the Class C airspace and draw close to the field, you're likely to be given a variety of instructions that will sequence you into the existing traffic flow. Whatever the messages, be sure to acknowledge and repeat them in an abbreviated form:

> *App: Cherokee Six One Tango, turn right heading three five zero. Descend and maintain four thousand.*

> **You:** Right to three five zero. Leaving six thousand five hundred for four thousand. Cherokee Six One Tango.

In another few minutes, you'll hear from Approach again:

> *App: Cherokee Six One Tango, Eppley is at twelve o'clock, six miles. Contact Eppley Tower on one two seven point six.*

> **You:** Roger. One two seven point six—and we have the field in sight. Cherokee Six One Tango. Good day.

Another handoff:

> **You:** Eppley Tower, Cherokee One Four Six One Tango with you, level at four thousand.

> *Twr: Cherokee Six One Tango, enter left downwind for Runway One Four left. Sequence later.*

Two points here:

1. "Sequence later" simply means the tower will advise you in due time whether you're cleared to land, are "number two following a Cessna on downwind" "number four following the Duchess," or whatever. Just remember to tell the tower that you "have the Cessna" or "negative contact on the Duchess" or "have the traffic." Always keep the tower informed—don't leave them in the dark.

2. You'll note that you've used the same squawk from Kansas City Center through Minneapolis Center, Omaha Approach, and Eppley Tower. No controlling agency has asked you to change. This is not always the case. Any one of them could have requested a different squawk and an ident. Leave the transponder on the current squawk until directed otherwise.

With a little time available now, dial in Eppley's Ground frequency, 121.9, on radio 1. Then, in due course, you'll hear on radio 2:

> *Twr: Cherokee Six One Tango, cleared to land.*

> **You:** Roger, cleared to land. Cherokee Six One Tango.

During the landing rollout, the tower makes its final contact with you:

> *Twr: Cherokee Six One Tango, contact Ground point niner clear of the runway.*

You: Cherokee Six One Tango, wilco.

Suppose, however, that you're not sure whether to make a left or right turnoff. You want to go to a Texaco dealer but don't know where one is located. Despite the uncertainty, don't tie up the tower frequency by asking the controller for directions. Let Ground do this for you, even if it means being cleared back across the active runway because you turned right instead of left, or vice versa. Merely acknowledge the tower's instructions and go to Ground's frequency:

You: Eppley Ground, Cherokee One Four Six One Tango Clear of One Four Left. Request progressive taxi to the Texaco dealer.

GC: *Cherokee One Four Six One Tango, turn right next taxiway. Taxi to One Four Right and hold for departing traffic.*

You: Roger, right and hold at One Four. Six One Tango.

After a minute or two:

GC: *Cherokee Six One Tango, Clear to cross One Four Right. Texaco is to your right at the Sky Harbor FBO.*

You: Roger, Ground. Clear to cross One Four—and we have the FBO. Six One Tango.

You're at the ramp, with the engine cut. Now don't forget to call Flight Service by phone or radio to close out your flight plan.

Omaha to Minneapolis

After an hour on the ground for refueling, a bite to eat, and a call to the Columbus, Nebraska, FSS for a weather briefing and filing the flight plan, you're ready to go again, with friends and baggage on board. The only new element is the need to call Clearance Delivery for initial VFR instructions within the Charlie airspace. As you've already determined, that frequency is 119.9, and you were told to contact Flight Service on the Omaha RCO frequency of 122.35. That's a change from the 122.2 you had listed on the frequency chart you prepared back in Kansas City (Fig. 14-3). So cross off 122.2 on that chart and enter 122.35 under the "Changes" column.

As usual, the communication chain begins by monitoring ATIS: "This is Eppley Information Gulf. One niner four five Zulu weather. Eight thousand scattered, visibility five, haze and smoke. Temperature eight five, dewpoint six two. Wind one four zero at one five. Altimeter three zero two five. ILS Runway One Four Right in use, land and depart Runway One Four Right. Advise you have Gulf."

Next, the call to Clearance (CD):

You: Eppley Clearance, Cherokee One Four Six One Tango.

CD: *Cherokee One Four Six One Tango, Eppley Clearance.*

You: Cherokee One Four Six One Tango will be departing Eppley, VFR northeast for Minneapolis at seven thousand five hundred.

CD: *Cherokee Six One Tango, Roger. Turn left heading zero four five after departure. Climb and maintain three thousand. Squawk two four four zero. Departure frequency one two four point five.*

You: Roger. Zero four five on the heading, maintain three thousand, two four four zero, and one two four point five. Cherokee Six One Tango.

CD: *Cherokee Six One Tango, readback correct.*

You: Cherokee Six One Tango.

Before calling Ground on radio 2, dial out the Clearance frequency in radio 1 and replace it with Flight Service's—122.35. Also, put the transponder on STANDBY.

You: Eppley Ground, Cherokee One Four Six One Tango at Sky Harbor with Information Gulf. Ready to taxi with clearance.

GC: *Cherokee Six One Tango, taxi to Runway One Four Right for intersection departure.*

You: Ground, Cherokee Six One Tango would like full length.

GC: *Cherokee Six One Tango, Roger. Full length approved.*

You: Cherokee Six One Tango.

After taxiing to the end of Runway 14, and with the pre-takeoff check completed, you next call Flight Service after advising Ground that you were going to change frequencies momentarily.

You: Columbus Radio, Cherokee One Four Six One Tango on one two two point three five, Eppley.

FSS: *Cherokee One Four Six One Tango, Columbus Radio.*

You: Columbus, Cherokee Six One Tango. Would you open my VFR flight plan to Minneapolis International at this time?

FSS: *Cherokee Six One Tango, Roger. We'll open your flight plan at two five.*

You: Roger, thank you. Cherokee Six One Tango.

That done, advise Ground Control that you're back with them. Now taxi to the hold and call the tower:

You: Eppley Tower, Cherokee One Four Six One Tango ready for takeoff, northeast departure.

Twr: *Cherokee Six One Tango, taxi into position and hold.*

You: Position and hold. Cherokee Six One Tango. [Switch the transponder from STANDBY to ALT as you're taxiing to the runway.]

Twr: *Cherokee Six One Tango, cleared for takeoff.*

You: Roger, cleared for takeoff. Cherokee Six One Tango. Is northeast departure approved?

Twr: *Affirmative, northeast departure approved.*

You: Roger. Cherokee Six One Tango.

When airborne, request the frequency change to Departure—if the tower has not already advised you to do so:

You: Tower, Cherokee Six One Tango requests frequency change to Departure.

Twr: *Cherokee Six One Tango, frequency change approved.*

You: Roger. Cherokee Six One Tango. Good day.

To Departure Control:

You: Omaha Departure, Cherokee One Four Six One Tango is with you out of one thousand eight hundred for three thousand. Request seven thousand five hundred.

Dep: *Cherokee Six One Tango, radar contact. Maintain present heading. Report level at three thousand.*

You: Roger, report three thousand, Cherokee Six One Tango.

You: Cherokee Six One Tango level at three thousand.

Dep: *Cherokee Six One Tango, Roger. Turn left heading zero three zero.*

You: Left to zero three zero. Cherokee Six One Tango.

Dep: *Cherokee Six One Tango, climb and maintain seven thousand five hundred.*

You: Roger. Out of three thousand for seven thousand five hundred. Cherokee Six One Tango.

You: Cherokee Six One Tango level at seven thousand five hundred.

Dep: *Cherokee Six One Tango, Roger. Proceed on course and contact Minneapolis Center on one two four point one.* [An unsolicited handoff for advisories.]

You: One two four point one. Thank you for your help. Cherokee Six One Tango.

To Center:

You: Minneapolis Center, Cherokee One Four Six One Tango with you, level at seven thousand five hundred.

Ctr: *Cherokee Six One Tango, Roger. Radar contact. Eppley altimeter Three Zero Two Five.*

You: Roger. Cherokee Six One Tango. Three Zero Two Five.

Down the road apiece, the summer turbulence at 7,500 feet is getting a bit too much for one of your passengers, so you decide to climb to 9,500. But first:

You: Minneapolis Center, Cherokee Six One Tango is out of seven thousand five hundred for niner thousand five hundred due to turbulence.

Ctr: *Cherokee Six One Tango, Roger. Report level at niner thousand five hundred.*

You: Wilco, Cherokee Six One Tango.

You: Cherokee Six One Tango level at niner thousand five hundred.

Ctr: *Cherokee Six One Tango, Roger.*

Just remember that on a VFR flight plan outside of a terminal airspace, you're free to deviate from existing headings and altitudes. But since you're asking Center for en-route advisories, don't make changes without advising the controller of your intentions. Although he or she can spot heading changes on the screen (and altitude changes, if you're equipped with a Mode C transponder), be sure to warn the controller in advance.

As you proceed toward Fort Dodge on V138, Center comes on again:

Ctr: *Cherokee Six One Tango, contact Minneapolis Center now on one three four point zero.* [This is the Fort Dodge remote outlet.]

You: Roger. One three four point zero. Cherokee Six One Tango. Good day.

Change the frequency accordingly and reestablish yourself with Center:

You: Minneapolis Center, Cherokee One Four Six One Tango with you level at niner thousand five hundred.

Ctr: *Cherokee Six One Tango, Roger. Fort Dodge altimeter three zero one six.* [Plus any instructions or traffic advisories the controller might have for you.]

Crossing the Fort Dodge VOR, you make a time check against your flight plan and find that you're running about 20 minutes behind schedule, which is quite a difference for the 110-mile flight from Eppley. Either the forecast winds have changed or those at the 9,500-foot altitude are from a different direction or velocity. Whatever the case, and because you intend to stay at the same altitude and generally northeast direction, it's fair to assume that your arrival in Minneapolis will be later than planned. With another 150 miles to go and at the present ground speed, you could easily be 45 minutes to an hour later than the flight plan ETA.

Before taking any arbitrary action, you decide that more information about the winds is in order, thus a call to Flight Service. First, however, advise Center if you're temporarily going to leave the frequency:

You: Center, Cherokee Six One Tango is leaving you temporarily to go to Flight Service.

> ***Ctr:*** *Cherokee Six One Tango, Roger. Advise when you're back with me.*

> **You:** Will do. Cherokee Six One Tango.

The FSS call is made on 122.3, the Fort Dodge transmit/receive frequency: (*Note*: the Fort Dodge FSS does not have Flight Watch service—which is why you make the call on 122.3 rather than 122.0.)

> **You:** Fort Dodge Radio, Cherokee One Four Six One Tango on one two two point three.

> ***FSS:*** *Cherokee One Four Six One Tango, Fort Dodge Radio, go ahead.*

> **You:** Fort Dodge, Cherokee Six One Tango is just north of the Fort Dodge VOR at niner thousand five hundred. Request winds at niner thousand feet.

> ***FSS:*** *Cherokee Six One Tango, Roger. Stand by.*

> ***FSS:*** *Cherokee Six One Tango, winds at niner thousand are three five zero at four zero. Fort Dodge altimeter two niner two five.*

> **You:** Three five zero at four zero and two niner two five. Thank you. Cherokee Six One Tango.

Without going through the mechanics of how you recompute your ground speed and ETA based on this information, let's just say that you determine that your arrival will be 48 minutes later than your flight plan forecast. This conclusion warrants another call to Flight Service:

> **You:** Fort Dodge Radio, Cherokee One Four Six One Tango on one two two point three.

> ***FSS:*** *Cherokee One Four Six One Tango, Fort Dodge Radio, go ahead.*

> **You:** Cherokee One Four Six One Tango is one zero north of the Fort Dodge VOR on VFR flight plan to Minneapolis International with a one six four five local ETA. Would like to extend the ETA to one seven three zero.

> ***FSS:*** *Cherokee Six One Tango, Roger. Will extend your ETA to one seven three zero local. Fort Dodge altimeter two niner two five.*

> **You:** Roger, thank you. Cherokee Six One Tango.

The next move is to go back to Center and reestablish contact:

> **You:** Center, Cherokee Six One Tango is back with you.

> ***Ctr:*** *Cherokee Six One Tango, Roger.*

With that taken care of, you can rest more easily. You have until 1800 before Flight Service will start asking questions as to your whereabouts. You've also allowed yourself an additional 27 minutes as an extra cushion.

Moving northward along V456, you'll be approaching another Center change point—this time to the remoted Mankato site. Somewhere near Mankato, you get this call:

Ctr: *Cherokee Six One Tango, contact Minneapolis Center now on one three two point four five.*

You: One three two point four five. Cherokee Six One Tango. Good day.

You: Minneapolis Center, Cherokee One Four Six One Tango with you, level at niner thousand five hundred.

Ctr: *Cherokee Six One Tango, Roger. Mankato altimeter two niner two three.*

You: Roger, two niner two three. Cherokee Six One Tango.

Immediately after passing the Mankato VOR, you leave the airway and turn to about 30 degrees in the direction of the Farmington VOR. Because you're only 50 miles from the airport at Mankato, however, it's almost certain that Center will turn you over to Approach and Approach will vector you to the airport area. That means you might not come anywhere near Farmington. Meanwhile, the closer you edge toward the Minneapolis Class B airspace, the greater the likelihood that Center will offer traffic advisories and possible heading changes. There might be none, but moving into a busy terminal environment increases the possibility.

About 20 miles out from the airspace limit, Center will conclude its radar surveillance. Perhaps the controller will hand you off to Approach. If so, the usual "with you," plus your altitude, is all that's necessary. If there is no handoff, though, you'll have to give the IPAI/DS—which means, among other things, knowing your position or approximate distance from the airport.

In this case; let's assume that Center is getting busy and doesn't have time to contact Approach. The controller comes on with:

Ctr: *Cherokee One Four Six One Tango, radar service terminated. Squawk one two zero zero. Contact Minneapolis Approach on one two five point zero. Descend your discretion.*

You: One two six point niner five. Will do. Cherokee Six One Tango.

"Descend your discretion" simply means what it says: You are clear to begin losing altitude whenever you wish and at whatever rate you wish. With that approval, you throttle back so that you'll be down to about 4,000 feet by the time you near the 30-mile Bravo airspace veil.

There's one more thing: Before calling Approach, dial in 120.8 for the current ATIS. Then go to 125.0, which is the Approach frequency for aircraft arriving from the south and west below 4500 feet, and make the introductory call:

You: Minneapolis Approach, Cherokee One Four Six One Tango.

App: *Cherokee One Four Six One Tango, Minneapolis Approach.*

You: Approach, Cherokee Six One Tango is thirty-five southwest over Le Center at five thousand three hundred descending for landing International and squawking one two zero zero with Delta.

App: *Cherokee Six One Tango, squawk four one two two and ident. Remain clear of the Bravo airspace.*

You: Roger, four one two two, and remaining clear. Cherokee Six One Tango.

App: *Cherokee One Four Six One Tango, radar contact thirty-two miles southwest. Cleared into the Minneapolis Bravo airspace. Turn left to zero two zero, descend and maintain four thousand five hundred.*

You: Roger, Cherokee Six One Tango cleared into Bravo airspace. Left to zero two zero and descending to four thousand five hundred.

You: Approach, Cherokee Six One Tango level at four thousand five hundred.

App: *Roger Six One Tango. Maintain heading and altitude.*

You: Will do. Six One Tango.

A short time later:

App: *Cherokee One Four Six One Tango, turn left to zero one zero and descend to two thousand five hundred.*

You: Roger, left to zero one zero and down to two thousand five hundred, Cherokee Six One Tango.

You: Approach, Cherokee Six One Tango level at two thousand five hundred.

App: *Roger, Six One Tango. Contact Minneapolis Tower on one two six point seven.*

You: One two six point seven. Will do, and thank you. Cherokee Six One Tango.

You: Minneapolis Tower, Cherokee One Four Six One Tango is with you, level at two thousand five hundred.

Twr: *Roger, Cherokee One Four Six One Tango. Descend to one thousand eight hundred for straight-in approach, landing Runway Four. Sequence later.*

You: Roger, descending to one thousand eight hundred and Runway Four. Cherokee Six One Tango.

You: Tower, Cherokee Six One Tango level at one thousand eight hundred.

Twr: *Roger, Six One Tango, you will be number two to land behind the Baron. Advise when you have traffic in sight.*

You: Number two, and will advise. Six One Tango.

You: Tower, Cherokee Six One Tango has the Baron.

> *Twr:* Roger Six One Tango. Thank you.

> *Twr:* Cherokee Six One Tango, cleared to land Runway Four.

> **You:** Cherokee Six One Tango, cleared to land.

When you're down:

> *Twr:* Cherokee Six One Tango, contact Ground point niner.

> **You:** Will do, tower. Six One Tango.

When you're clear of the active and at a full stop:

> **You:** Minneapolis Ground, Cherokee One Four Six One Tango clear of Four. Taxi to Signature.

> *GC:* Roger. Cherokee One Four Six One Tango, taxi to Signature.

Once again, when you're parked or in the pilot's lounge, close out the flight plan with Flight Service—which, in this case, is located in Princeton, Minnesota.

Minneapolis to Mason City

It's the next morning and you're ready to set out for Mason City, Newton, and then back to Kansas City. First the normal routines: checking the weather, filing the flight plan, and determining the FSS frequency to open the flight plan. With the engine started, monitor the Minneapolis departure ATIS on 135.35. Remember that you're in a Class B airspace, so the first call goes to Clearance Delivery on 133.2:

> **You:** Minneapolis Clearance, Cherokee One Four Six One Tango.

> *CD:* Cherokee One Four Six One Tango, Clearance.

> **You:** Cherokee One Four Six One Tango will be departing VFR to Mason City. Request seven thousand five hundred.

> *CD:* Cherokee Six One Tango, cleared into the Bravo airspace. Turn right heading one seven five after departure. Climb and maintain three thousand. Squawk zero four two zero. Departure frequency one two four point seven.

> **You:** Cherokee Six One Tango, cleared into the Bravo airspace, right heading one seven five after departure, maintain three thousand, zero four two zero, and one two four point seven.

> *CD:* Cherokee Six One Tango, readback correct.

> **You:** Roger. Cherokee Six One Tango.

Now to Ground Control on 121.9:

> **You:** Minneapolis Ground, Cherokee One Four Six One Tango at Airmotive with Information Echo and clearance.

> *GC:* Cherokee Six One Tango, taxi to Runway One One Right.

You: Roger, One One Right, Cherokee Six One Tango.

The pre-takeoff check has been completed. Before taxiing to the hold line, ask Ground for permission to switch frequencies, then call Flight Service to open the flight plan.

You: Princeton Radio, Cherokee One Four Six One Tango on one two two point five five [the RCO frequency FSS gave you when filing], Minneapolis.

FSS: *Cherokee One Four Six Tango, Princeton Radio.*

You: Roger, would you please open my flight plan to Mason City at 1410 UTC?

(*Note*: It is now 0900 Central Daylight Time, or 1400 UTC Greenwich Time. You have added 10 minutes to your departure time in case of any delay.)

FSS: *Cherokee Six One Tango, Roger. Will open your flight plan at one zero.*

You: Thank you. Cherokee Six One Tango.

As you move toward the hold line, you see that two aircraft are ahead of you awaiting takeoff permission. Regardless, you pull behind the second plane and call the tower on 126.7:

You: Minneapolis Tower, Cherokee One Four Six One Tango ready for takeoff, number three in sequence, south departure.

Twr: *Cherokee One Four Six One Tango, taxi around the Mooney and Skymaster. Cleared for takeoff, south departure approved.*

You: Roger, cleared for takeoff, Cherokee Six One Tango.

At this point, switch the transponder from SBY to ALT. When airborne, turn to your assigned heading of 175 degrees. The tower will probably authorize the frequency change to Departure, but if it doesn't, request the change:

You: Tower, Cherokee Six One Tango requests frequency change to Departure.

Twr: *Cherokee Six One Tango, frequency change approved. Good day.*

You: Roger. Cherokee Six One Tango. Good day.

Switch to your other radio, already dialed in to 124.7:

You: Minneapolis Departure, Cherokee One Four Six One Tango with you out of one thousand seven hundred for three thousand.

Dep: *Cherokee One Four Six One Tango, report level at three thousand.*

You: Wilco, Cherokee Six One Tango.

You: Cherokee Six One Tango level at three thousand.

Dep: *Cherokee Six One Tango, Roger.*

Dep: *Cherokee Six One Tango, climb and maintain four thousand five hundred.*

You: Roger. Out of three thousand for four thousand five hundred. Cherokee Six One Tango.

You: Cherokee Six One Tango level at four thousand five hundred.

Dep: *Cherokee Six One Tango, Roger. Cleared direct Farmington VOR.*

You: Roger, cleared direct Farmington. Cherokee Six One Tango.

Dep: *Cherokee Six One Tango, climb and maintain seven thousand five hundred.*

You: Roger. Out of four thousand five hundred for seven thousand five hundred. Cherokee Six One Tango.

You: Cherokee Six One Tango level at seven thousand five hundred.

Dep: *Cherokee Six One Tango. Roger. Report Farmington VOR.*

You: Roger, report Farmington. Cherokee Six One Tango.

It isn't long before the Course Direction Indicator (CDI) swings erratically and the VOR flag changes from TO to FROM. You're over the Farmington VOR. You report in and are cleared to turn to the 178-degree heading which establishes you on Victor 13. As you're still in the Bravo airspace, other instructions might be forthcoming from Departure. In a very few minutes, you'll hear something like this:

Dep: *Cherokee Six One Tango, position fifteen miles south of Farmington VOR, departing the Bravo airspace. Radar service terminated. Squawk one two Zero Zero. Frequency change approved. Good day.*

You: Departure, Cherokee Six One Tango. Can you hand us off to Center?

Dep: *Cherokee Six One Tango, stand by. [Pause] Cherokee Six One Tango, unable at this time. Squawk one two zero zero. Suggest you contact Minneapolis Center on one three four point eight five.*

You: Roger. One two zero zero and one three four point eight five. Cherokee Six One Tango.

You: Minneapolis Center, Cherokee One Four Six One Tango.

Ctr: *Cherokee One Four Six One Tango, Minneapolis Center.*

You: Cherokee One Four Six One Tango is fifteen south of the Farmington VOR at seven thousand five hundred, VFR to Mason City, squawking one two zero zero. Request VFR advisories, if possible.

This time, the press of traffic in and out of the Minneapolis area is of such density that Center can't accept your request:

Ctr: *Cherokee One Four Six One Tango, unable at this time. Suggest you monitor this frequency.*

You: Roger, Center. Understand. Cherokee Six One Tango.

With only 80 miles or so to go, this isn't much of a problem. However, it does mean that the need for constant sky-scanning is more important than ever. If Center is too busy to give you advisories, you can be reasonably certain that there's a fair amount of activity in the surrounding area. Maximum alertness is thus in order.

As you've already noted in your flight planning, Mason City is a non-tower-controlled Class E airport with only unicom on 123.0 available for airport advisories. The Fort Dodge FSS can be reached on the 122.6 RCO frequency for opening and closing flight plans. First things first, though, so you begin letting down about 15 miles north of the airport and call unicom on its CTAF:

You: Mason City Unicom, Cherokee One Four Six One Tango.

Uni: *Cherokee One Four Six One Tango, Mason City Unicom.*

You: Unicom, Cherokee One Four Six One Tango is fifteen north at five thousand three hundred descending for landing. Request field advisory.

Uni: *Six One Tango, wind is two zero zero at one zero, altimeter two niner four five. Favored runway is One Seven. No reported traffic.*

You: Roger, thank you. Six One Tango.

From this point on, remember to address all flight operation reports to "Mason City Traffic," not unicom. If, however, your deplaning passenger wants a cab or a telephone call made, or you have a request of a nonoperational nature, the message is addressed to "Mason City Unicom."

Even though there is no "reported" traffic and a straight-in approach to Runway 17 would be the fastest way to get on the ground, you're aware that non-tower-controlled airport pattern procedures call for standard downwind and base legs. Consequently, your next call to Mason City traffic goes like this:

You: Mason City Traffic, Cherokee One Four Six One Tango is 10 miles north at three thousand. Will cross over the airport to the southeast at three thousand for one eighty and entry to left downwind, landing Mason City.

A few minutes later, as you complete the 45-degree entry to the downwind, you report your position:

You: Mason City Traffic, Cherokee Six One Tango entering left downwind at midfield for landing Runway One Seven, Mason City.

Then, as you're turning to base and final:

You: Mason City Traffic, Cherokee Six One Tango turning left base for landing One Seven, Mason City.

You: Mason City Traffic, Cherokee Six One Tango turning final, landing Mason City.

Finally, when down and clear of the runway:

You: Mason City Traffic, Cherokee Six One Tango clear of One Seven, taxiing to the ramp, Mason City.

At the ramp you decide to close the flight plan by radio rather than by telephone. Using the RCO you call the FSS:

You: Fort Dodge Radio, Cherokee One Four Six One Tango on one two two point six, Mason City.

FSS: *Cherokee One Four Six One Tango, Fort Dodge Radio, go ahead.*

You: Cherokee Six One Tango is on the ramp at Mason City. Would you close out my VFR flight plan from Minneapolis at this time?

FSS: *Cherokee Six One Tango, Roger. Will close out your flight plan at five five.*

You: Thank you. Cherokee Six One Tango.

Mason City to Newton

The flight plan to Newton, 90 miles away, has been filed, you know the weather, and it's departure time again. Tune to 123.0 and announce your initial intentions:

You: Mason City Traffic, Cherokee One Four Six One Tango at the terminal, taxiing to Runway One Seven, Mason City.

After the pretakeoff check, with the second radio already turned to 122.6, open the flight plan:

You: Fort Dodge Radio, Cherokee One Four Six One Tango on one two two point six, Mason City.

FSS: *Cherokee One Four Six One Tango, Fort Dodge Radio, go ahead.*

You: Fort Dodge, would you please open my flight plan to Newton at this time?

FSS: *Cherokee Six One Tango, Roger. Will open your flight plan to Newton at two zero.*

You: Roger. Thank you. Cherokee Six One Tango.

Now go back to 123.0:

You: Mason City Traffic, Cherokee Six One Tango taking One Seven, straight-out departure, Mason City.

When clear of the Class E (Echo) surface area:

You: Mason City Traffic, Cherokee Six One Tango is clear of the area to the south, Mason City.

Following this call, you request Minneapolis Center to give you VFR advisories. Center in this location is remoted to Mason City on 127.3:

You: Minneapolis Center, Cherokee One Four Six One Tango.

Ctr: *Cherokee One Four Six One Tango, Minneapolis Center, go ahead.*

You: Center Cherokee One Four Six One Tango is off Mason City at four thousand five hundred, climbing to seven thousand five hundred, enroute Newton, and squawking one two zero zero. Request VFR advisories.

Ctr: *Cherokee Six One Tango, squawk two five two five and ident.*

You: Cherokee Six One Tango, two five two five.

Ctr: *Cherokee Six One Tango, radar contact. Traffic at ten o'clock, four miles, southbound. Altitude unknown.*

You: Cherokee Six One Tango is looking.

A minute or so later, you spot the target a little above you at the eleven o'clock position:

You: Cherokee Six One Tango has the traffic.

Ctr: *Cherokee Six One Tango, Roger.*

When at your altitude:

You: Center, Cherokee Six One Tango level at seven thousand five hundred.

Ctr: *Cherokee Six One Tango, Roger.*

Very shortly, according to the Enroute Low Altitude Chart (ELAC), you'll be leaving the airspace of Minneapolis Center and entering that controlled by Chicago. You can't be sure, but you'll probably be asked to change to the Des Moines remote outlet on 127.05. Assuming that will be the frequency, dial it in so that you'll be prepared. After a few more minutes, Center comes on:

Ctr: *Cherokee Six One Tango, contact Chicago Center now on one two seven point zero five. Good day.*

You: Roger, one two seven point zero five. Thank you for your help. Cherokee Six One Tango.

You: Chicago Center, Cherokee One Four Six One Tango is with you, level at seven thousand five hundred.

Ctr: *Cherokee Six One Tango, radar contact. Des Moines altimeter two niner niner eight.*

Nearing Newton, you'll hear something like this:

Ctr: *Cherokee Six One Tango, position one five miles north of the Newton VOR. Radar service terminated. Squawk one two zero zero. Frequency change approved. Good day.*

You: Roger. One two zero zero. Thank you for your help. Cherokee Six One Tango.

Newton is another non-tower-controlled Class E Airport, with only unicom on 122.8. About 10 miles out, you announce your presence:

> **You:** Newton Unicom, Cherokee One Four Six One Tango is five north of Newton VOR. Request airport advisory.
>
> *Unicom:* *Cherokee One Four Six One Tango, Newton Unicom. Wind is two one zero at one five. Altimeter three zero one five. Runway One Three in use. Three Cessnas reported in the pattern.*
>
> **You:** Roger. Cherokee Six One Tango.

This is another situation where your position and the runway-in-use invite a straight-in approach and landing. Again, though, you intend to go by the rules and adhere to the standard traffic pattern procedures. Hence this call:

> **You:** Newton Traffic, Cherokee One Four Six One Tango is eight miles northwest at four thousand, descending. Will cross over the airport at two thousand five hundred for standard entry to downwind for landing One Three Newton.

While in the process of turning to the downwind leg from the 45-degree entry, your next calls follow the routine non-tower-controlled Class E or G format:

> **You:** Newton Traffic, Cherokee Six One Tango entering left downwind at midfield for landing One Three Newton.
>
> **You:** Newton Traffic, Cherokee Six One Tango turning left base for landing One Three Newton.
>
> **You:** Newton Traffic, Cherokee Six One Tango turning final for landing One Three Newton.

And when down and clear:

> **You:** Newton Traffic, Cherokee Six One Tango clear of One Three. Taxiing to the terminal, Newton.

When in the terminal, don't forget to cancel the flight plan. At Newton, this has to be done by phone to the FSS in Fort Dodge on 1-800-WX-BRIEF.

Newton to Kansas City

With the remaining passenger dropped off and full fuel tanks, you're ready for the last leg back to Kansas City. But first comes another call to Flight Service for a weather check and flight plan filing. Local conditions are determined from the unicom operator (or review the winds, etc., yourself if he's out gassing an airplane).

The radio contacts will first be to local traffic and then to Flight Service, which you won't be able to reach until you have some altitude. Heading southwest to the Des Moines VOR, you'll be well above the 5,000 foot msl ceiling of the Des Moines Class C, so calling Approach isn't necessary. From the VOR southbound on V13, however, you'll be in the Class C airspace's outer area, which rises to approximately 12,000 feet msl, for a few minutes. Even though the weather is fine, you conclude that monitoring Approach would be a good idea, and, if the volume of traffic indicates, asking for advisories.

While you're still on the ramp at Newton:

You: Newton Traffic, Cherokee One Four Six One Tango at the terminal, taxiing to Runway One Three, Newton.

After engine runup:

You: Newton Traffic, Cherokee One Four Six One Tango taking One Three, southwest departure, Newton.

Off the ground and at 300 feet or so, you contact Fort Dodge Flight Service over the Newton VOR. In this instance, you transmit on 122.1 and receive on the VOR frequency of 112.5.

You: Fort Dodge Radio, Cherokee One Four Six One Tango listening Newton VOR.

FSS: *Cherokee One Four Six One Tango, Fort Dodge Radio, go ahead.*

You: Fort Dodge, Cherokee Six One Tango was off Newton at three five past the hour. Would you please open my flight plan to Kansas City Downtown?

FSS: *Cherokee Six One Tango, Roger. We show you off Newton at three five and will activate your flight plan to Kansas City. Des Moines altimeter three zero zero one.*

You: Roger. Thank you. Cherokee Six One Tango.

Finally, one more local call is in order:

You: Newton Traffic, Cherokee Six One Tango now clear of the area to the southwest, Newton.

Monitoring Des Moines Approach as you pass through the Class C airspace's outer area, you call Minneapolis Center when clear of the area.

You: Minneapolis Center, Cherokee One Four Six One Tango.

Ctr: *Cherokee One Four Six One Tango, Minneapolis Center, go ahead.*

You: Center, Cherokee One Four Six One Tango is fifteen southwest of the Newton VOR at five thousand two hundred, climbing to six thousand five hundred, VFR to Kansas City Downtown via Victor One Three. Squawking one two zero zero. Request VFR advisories, workload permitting.

Ctr: *Cherokee Six One Tango, squawk zero five two three and ident.*

You: Cherokee Six One Tango squawking zero five two three.

Ctr: *Cherokee Six One Tango, radar contact. Report level at six thousand five hundred.*

Ctr: *Cherokee Six One Tango, Roger.*

There might or might not be further advisories from Center, depending on traffic. Whichever the case, you're soon over the Des Moines VOR and heading outbound on Victor 13.

After passing the Lamoni VOR, 53 miles out of Des Moines, you leave Minneapolis Center and enter the Kansas City Center area. As you cross the line:

Ctr: *Cherokee Six One Tango. Contact Kansas City Center now on one two seven point niner. Good day.*

You: Roger. One two seven point niner. Cherokee Six One Tango. Good day.

You: Kansas City Center, Cherokee One Four Six One Tango is with you at six thousand five hundred.

Ctr: *Cherokee Six One Tango, Roger. St. Joe altimeter two niner niner six.*

You: Roger, Cherokee Six One Tango.

About now is the time to tune the second radio to the Kansas City VOR on 112.6. With one VOR head tracking you outbound from Lamoni and the other inbound to Kansas City, you should be smack in the middle of V13.

When you're approximately due east of St. Joseph, Missouri, you spot some lightning not too far distant and just to the right of your course. This can be an omen of bad stuff, so you decide to check with Center.

You: Center, Cherokee Six One Tango. Request.

Ctr: *Cherokee Six One Tango, go ahead.*

You: Cherokee Six One Tango has lightning at one o'clock. Will my present course keep me clear of the storms?

Ctr: *Cherokee Six One Tango, affirmative. Scattered thunderstorms are moving northeast, but you should be past the area at your present ground speed.*

You: Roger. Thank you. Cherokee Six One Tango.

On the other hand, if a storm encounter seems likely, Center might offer this advice: "Cherokee Six One Tango, the storm activity is moving due east. Suggest right heading of two seven zero past St. Joe to Topeka and come in behind the weather." Keep in mind that such a suggestion is not a command. You're VFR, and have freedom as well as options.

Assuming the weather is not going to be a factor, Center will call you as you near the Kansas City TCA:

Ctr: *Cherokee Six One Tango, position one five miles north of the Kansas City Bravo airspace. Contact Kansas City Approach on one one niner point zero.*

You: Roger, one one niner point zero. Thank you for your help. Cherokee Six One Tango.

Before contacting Approach, be sure you have monitored the Kansas City Downtown ATIS. Then, with the information clearly in mind, call Approach:

You: Kansas City Approach, Cherokee One Four Six One Tango is with you, level at six thousand five hundred with Foxtrot.

App: *Cherokee One Four Six One Tango, cleared into the Bravo airspace, direct Kansas City VOR. Descend and maintain three thousand. Remain VFR at all times.*

You: Cherokee Six One Tango, Roger. Cleared into the Bravo airspace, direct Kansas City VOR, down to three thousand and remain VFR.

You: Approach, Cherokee Six One Tango level at three thousand.

App: *Roger, Six One Tango.*

Approach sees you nearing the VOR:

App: *Cherokee Six One Tango, descend to two thousand five hundred.*

You: Cherokee Six One Tango out of three for two thousand five hundred.

You: Approach, Cherokee Six One Tango level at two thousand five hundred.

App: *Roger, Six One Tango. Maintain heading and altitude.*

You: Will do. Six One Tango.

As you cross the VOR, only 9 miles from the airport:

App: *Cherokee Six One Tango, Downtown is niner miles at twelve o'clock. Advise when you have the airport in sight.*

You: Approach, Six One Tango has the airport.

App: *Roger, Six One Tango. Contact Downtown Tower on one three three point three.*

You: Will do. Cherokee Six One Tango.

You: Downtown Tower, Cherokee One Four Six One Tango is with you, level at two thousand five hundred.

Twr: *Cherokee One Four Six One Tango, continue straight in for Runway One Niner. Sequence later.*

You: Roger, straight in for One Niner, Cherokee Six One Tango.

Twr: *Cherokee Six One Tango, you'll be number two to land following a Citation on base.*

You: Roger. Cherokee Six One Tango has the Citation.

Twr: *Cherokee Six One Tango, Roger. Caution wake turbulence landing Citation. Cleared to land Runway One Niner.*

You: Cleared to land, Cherokee Six One Tango.

You watch the Citation touch down. To plan your landing because of the wake turbulence, you'd like a current wind reading.

You: Tower, Cherokee Six One Tango. Wind check.

Twr: *Cherokee Six One Tango, wind two one zero at one five.*

You: Cherokee Six One Tango.

Because you're coming in on Runway 19, these winds should blow the wake to the left of the runway, so you decide to land on the right, or upwind, side. You do so without difficulty and complete the rollout.

Twr: *Cherokee Six One Tango, contact Ground point niner when clear.*

You: Cherokee Six One Tango.

You: Downtown Ground, Cherokee One Four Six One Tango clear of One Niner. Taxi to Hangar Six.

GC: *Cherokee Six One Tango, taxi to Hangar Six.*

By radio when parked, or by phone, you close out the flight plan—and the trip concludes without incident.

CONCLUSION

The point of this cross-tower country was to illustrate the typical radio procedures when using Clearance Delivery Contacting and Ground Control, Tower, Approach/Departure, Center, Flight Service, when operating within Class B, C, D, and E airspaces and airports. The trip tried to encapsulate the more common phrases and phraseologies discussed in the various previous chapters.

Yes, there were several instances of what you might have considered needless repetition, but a cross-country involves repetition. Many of the same things are said to different agencies. Besides, repetition has a way of cementing habits in our minds and preventing errors and misunderstandings. Not every possible contact was included (as, for example, a call to Flight Watch), but many of the most common dialogues were re-created.

And, yes, some of the dialogues might seem a little stilted. As I stated early in the book, however, the examples are based on those illustrated in the *Aeronautical Information Manual* (*AIM*) and established as policy in the controller's *Air Traffic Control* manual, 7110.65G.

If you fly enough, you'll occasionally hear minor variations or local adaptations, particularly on the part of pilots. It's probably inevitable that a little slang or jargon, neither of which necessarily distorts the message, creeps in. Such liberties, however, don't conform to FAA procedures. Consequently, I've tried to be as accurate and literal as possible in illustrating the approved communicating techniques—both in this theoretical cross-country and throughout the entire book.

In the process, I hope that the use of the radio and the services available to every pilot are just a bit clearer. If such is the case, perhaps some of the communicating concerns experienced by so many VFR pilots—both new and experienced—have been allayed, at least a little.

17
A Final Word

PROPER RADIO PROCEDURES IS PERHAPS THE MOST OVERLOOKED OR underemphasized subject in pilot training programs. Many budding pilots are never taught how to record and organize nav and com frequencies for easy reference and how to set up the frequencies in advance (assuming two navcoms are on board) to avoid excessive dialing at changeover points. They receive only the very basics of air-to-ground communications, including what to say to whom, when to say it, how to say it, and what to expect to hear in response.

Of course, there are exceptions to these criticisms. Some instructors like to teach communications and produce excellent pilots in the process. Their students sally forth into controlled areas with skill and confidence.

For those not privy to such training, it's a different story. Even many "experienced" pilots fear that they won't understand an instruction from a controller and that they will sound stupid over the air. The result is that those pilots avoid controlled airports and other FAA services available to them—services already paid for by their tax dollars. The fears are both natural and understandable for pilots who haven't been trained in radio techniques.

Everyone has qualms when they make those first tentative calls. Everyone has screwed up a transmission at one time or another. So what? Mistakes in flying an airplane can be very fatal; mistakes over the air are usually no more than embarrassing—if that. If you learn what to say, how to say it, and aren't afraid to ask a controller to "say again" or "say more slowly" when you haven't understood, you'll find that your flights, local or otherwise, will be safer and more secure, and you'll have the confidence to venture into that tower-controlled airport you perhaps have been avoiding.

If this book has clarified just a few of the areas in the radio communications process, it has met its objective. The project was undertaken because the need existed for a work of this nature. By so doing, I hope that any concerns have been put to rest and that your flying will be more enjoyable.

There are others in the air, however, who have no concerns and no doubts, but are languishing in unrecognized ignorance. They bust into controlled airspaces unannounced; they monopolize the air with trivia; they hem, haw, mumble, and meander; their messages are unplanned, their transmissions disorganized. They take four minutes to communicate what the pro does in four seconds. Are they aware of their deficiencies? You know better. The air is for them, and let the rest take the hindmost . . . .. As is so often the case, those who need help the most are the last to recognize that need and ask for the help.

The sky is for all of us. For those who like to venture forth, do it with confidence and professionalism. Radio skills don't make better pilots at the controls, but they certainly add to competence and, quite logically, professionalism. Therein lies much of the joy of flying.

Abbreviations and Acronyms

The following is a glossary of various terms, abbreviations, and acronyms that appear in this book. For a more complete glossary of pilot/controller radio communication terms, refer to the *Aeronautical Information Manual*.

AAS	Airport Advisory Service
ADF	Automatic Direction Finder
A/FD	Airport/Facility Directory
A/G	Air/Ground Communications
agl	Above ground level
AIM	*Aeronautical Information Manual*
AIRMET	Airman's Meteorological Information
ALT	Transponder switch position to activate altitude-reporting feature
AOPA	Aircraft Owner and Pilots Association
App	Approach Control
ARTCC	Air Route Traffic Control Center
ARTS	Automated Radar Terminal System
ASOS	Automated Surface Observation System
ATC	Air Traffic Control

Abbreviations and acronyms

ATCRBS	Air Traffic Control Radar Beacon System
ATIS	Automatic Terminal Information System
ATP	Air Transport Pilot
AWOS	Automated Weather Observing System
CAVU	Ceiling And Visibility Unlimited
CD	Clearance Delivery (*also see* **CLNC DEL**)
Center	Air Route Traffic Control Center (*also see* **ARTCC**)
CFI	Certified Flight Instructor
CLNC DEL	Clearance Delivery (*also see* **CD**)
Com	Communications (also, send-and-receive side of the radio or navcom)
CT	Control Tower
CTAF	Common Traffic Advisory Frequency
Dep	Departure Control
DF	Direction-finding
DME	Distance Measuring Equipment
EFAS	Enroute Flight Advisory Service (Flight Watch)
ELAC	Enroute Low Altitude Chart
ELT	Emergency Locator Transmitter
ETA	Estimated Time Enroute
ETD	Estimated Time of Departure
ETE	Estimated Time Enroute
FAA	Federal Aviation Administration
FARs	Federal Aviation Regulations
FBO	Fixed-base Operator
FCT	Federal Control Tower
FSS	Flight Service Station
GC	Ground Control (*also see* **GND CON**)
GND CON	Ground Control
HIWAS	Hazardous Inflight Weather Advisory Service
ICAO	International Civil Aviation Organization
IMC	Instrument Meteorological Conditions
IPAI/DS	Identification-Position-Altitude-Intentions or Destination-Squawk
IR	Military Instrument Flight Route
MEA	Minimum Enroute Altitude
MOA	Military Operations Area
MOCA	Minimum Obstruction Clearance Area
MRA	Minimum Reception Area
msl	Mean sea level
MTR	Military Training Route
multicom	Nongovernment air-to-air radio communication frequency
Nav	Navigation (also the navigation side of the radio, or "navcom")
NDB	Nondirectional Beacon
NF	Non-Federal Control Tower (*also see* **NFCT**)

NFCT	Non-Federal Control Tower
NOTAM	Notice To Airmen
NTSB	National Transportation Safety Board
PIREP	Pilot (weather) report
RAPCON	Radar Approach Control (military)
RCO	Remote Communications Outlet
RDT&E	Research, Development, Testing, and Evaluation
RF	Radio Failure
SAR	Search-and-Rescue
SBY	Standby (a transponder switch position)
SFAR	Special Federal Aviation Regulation
SIGMET	Significant Meteorological Information
sm	Statute mile
Squawk	Reference to transmission of an assigned transponder code
SUA	Special Use Airspace
SVFR	Special VFR rules or operations
TAC	Terminal Area Chart
TACAN	Tactical Air Navigation
TRACAB	Terminal Radar Control in the Tower Cab
TRACON	Terminal Radar Control
TRSA	Terminal Radar Service Area
TWEB	Transcribed Weather Broadcast
UHF	Ultrahigh Frequency
unicom	Nongovernment air-to-ground radio communication facility
UTC	Coordinated Universal Time (Greenwich Mean Time [GMT]) (*also see* **ZULU**)
VFR	Visual Flight Rules
VHF	Very High Frequency
VOR	Very high frequency Omnidirectional Range (for course guidance information)
VORTAC	A VOR transmitter combined with military TACAN distance information
VR	Military Visual Flight Rules Route
Wilco	"Will comply"
XFSS	Auxiliary Flight Service Station
ZULU	UTC time frequently cited as "_____ Zulu time"

Index

Index

Index

N

Nall Report (*See also* accidents due to poor communications), 12

National Transportation Safety Board (NTSB), 12, 13

near mid-air collision, 14–15

N-numbers, aircraft identification (*see* call-signs)

nondirectional radio beacons (NDB), 56–57

Notices to Airmen (NOTAM), 66

O

outer area, Class C airspace, 38, 165, 168–169

outer circle, Class C airspace, 38

P

phonetic designations, ATIS, 101

phraseology, 8–9

pilot reports (PIREPS), 46, 66, 81, 82

poor communications example, 1–3, 13–24

position reporting, flight service stations (FSS), 93–95

positive control area (PCA) (*see* Class A airspace)

practicing radio communications, 5–6

precipitation, 81

preflight briefings, 67–71

prohibited airspace, 40

R

radio checks, ground control communication, 108

radio failure, 219–224

7600 transponder code, 221–222

informing control, 220–224

light gun signals, 221–222

preventing failures, 219–220

testing the radio, 220

Rapcon Approach/Departure control, 178

remote communications outlet (RCO), 66, 71–72

opening flight plan by radio, 75–76

repeating the message, 32–33

restricted airspace, 40–41, 66

route of flight, 226–227

S

separation minima, 42

Class C airspace, 167

Class D airspace, 144–145

shyness, 31–32

SIGMETs, 80–81

special use airspace (SUA), 40–42, 84–85

special VFR (SVFR), 86–89, 129

Class D airspace, 148–149

speed of thought vs. speech, 28–29

SQUAWK codes, transponder, 117–118

standard briefing, 68

T

TACAN, 38

takeoff/departure, 233–234, 239–242

airport underlying Class B airspace, 182–184

Approach/Departure Control, 177–198

Class B airspace, 161–162, 179–182, 246–248

Class C airspace, 184–186

Class D airspace, 186–187

Class D airspace, tower communication, 136–138

Class E airspace, 250

frequency change at takeoff, 138–139

ground control communications, 104–107

handoff to ARTCC, 207–210

multicom takeoff, sample communication, 51–53

satellite airports, Class C, 186

terminal radar service area (TRSA), 173

touch-and-go, 139

touch-and-go, multicom, 53

touch-and-go, unicom, 62–63

unicom, sample communication, 61–63

taxiing, taxi out/taxi back

flight service station (FSS) information, 91

ground control communications, 104–107

progressive taxi instructions, 106–107

terminal area charts, 38

terminal control areas (TCA) (*see* Class B airspace)

terminal radar service area (TRSA), 39–40, 154, 170–173, 174

Airport/Facilities Directories listing, 170

approach/landing contact, 173

chart symbols for, 170, 171

establishing contact, 170–172

services provided within, 170–173

Index